AQA Geography

Revision Guide

A LEVEL & AS

HUMAN GEOGRAPHY

Series editor   Alice Griffiths

Lawrence Collins

Tim Bayliss

Simon Ross

OXFORD
UNIVERSITY PRESS

OXFORD
UNIVERSITY PRESS

Great Clarendon Street, Oxford, OX2 6DP, United Kingdom

Oxford University Press is a department of the University of Oxford.
It furthers the University's objective of excellence in research, scholarship,
and education by publishing worldwide. Oxford is a registered trade mark
of Oxford University Press in the UK and in certain other countries

© Oxford University Press 2018

Series editor: Alice Griffiths

Authors: Lawrence Collins, Tim Bayliss, Simon Ross

The moral rights of the authors have been asserted.

Database right of Oxford University Press (maker) 2018.

First published in 2018

British Library Cataloguing in Publication Data

Data available

ISBN 978-019-843269-2

10 9 8 7 6 5 4 3 2 1

Printed in Italy by L.E.G.O. SpA

Acknowledgements

The publisher and authors would like to thank the following for permission
to use photographs and other copyright material:

Cover: © Patrick Bingham-Hall; **p8(t):** Ralph Hagen/Cartoonstock;
p8(b): Randy Glasbergen; **p18:** Shutterstock; **p28:** The Coca Cola Case
©2009 Argus Film/National Film Board of Canada. All rights reserved;
p29: Shutterstock; **p30:** Eric St Pierre/Equal Exchange Coop; **p31:** Joki
Desnommée-Gauthier; **p33:** ZUMA Press, Inc./Alamy Stock Photo; **p34**(t,
b): Shutterstock; **p35, 36(t):** Shutterstock; **p36(b):** Global Goals/www.
globalgoals.org; **p37(t, b):** Shutterstock; **p38, 39:** Shutterstock; **p42:**
Professor Michael Ashley/University of New South Wales; **p43:** popa/
Toonpool; **p46:** Rudi Van Starrex/Getty Images; **p48:** YHA (England and
Wales); **p49:** Shutterstock; **p51(l):** allan wright/Alamy Stock Photo; **p51(r):**
Jack Sullivan/Alamy Stock Photo; **p52:** paul weston/Alamy Stock Photo;
p53(t): 1987: REUTERS/Stringer, 2013: REUTERS/Carlos Barria; **p53(b),
54:** Shutterstock; **p55:** Courtesy of Alice Liddell Innovative Community
Enterprise Ltd; **p56:** Courtesy of Visit Britain; **p57:** Shutterstock; **p58:**
Datashine/James Cheshire and Oliver O Brien/UCL Geography. Place names
and buildings: Ordinance Survey © Crown copyright and Database rights
2014-15.; **p59(t):** Department for Communities and Local Government
(OGL); **p59(b):** © Consumer Data Research Centre 2016/© Crown copyright
& database right 2014-5; **p60:** Jeff Gilbert / Alamy Stock Photo; **p61:**
Crown Copyright (2016)100043706/Chilterns Conservation Board; **p62:**
www.littletoller.co.uk; **p63:** Roy Porter; **p65(t):** Kristoffer Tripplaar/Alamy
Stock Photo; **p65(b):** Jim West/Alamy Stock Photo; **p67:** Eric Fischer/©
OpenStreetMap, CC-BY-SA; **p68:** © 2014 (Alana Semuels/Los Angeles Times/
MCT) All rights reserved. Distributed by Tribune Content Agency; **p75:**
Shutterstock; **p78:** Bluesky International Limited; **p80:** Tim Bayliss; **p83(t):**
Shutterstock; **p83(b):** Tim Bayliss; **p84:** Shutterstock; **p85(t):** "Changing
Tastes in Britain stamps" designed by Catell Ronca and Rose Design ©
Royal Mail Group Limited, 2005; **p85(b):** Matthew Lewis/Getty Images;
p89: Walter Bibikow/Getty Images; **p91:** OUP; **p92:** Royal Haskoning; **p93:**
Airfotos/Northumbrian Water; **p95:** Aurora Photos/Alamy Stock Photo;
p97(t): Tim Bayliss; **p97(b):** Chris Cooper-Smith/Alamy Stock Photo; **p98:**
Shutterstock; **p99:** imageBROKER/Alamy Stock Photo; **p100:** Shutterstock;
p101(t): Peter M. Wilso /Alamy Stock Photo; **p101(b):** By Halley Pacheco
de Oliveira (Own work) [CC BY-SA 3.0 (http://creativecommons.org/
licenses/by-sa/3.0)], via Wikimedia Commons; **p102:** Shutterstock; **p108:**
Tim Bayliss; **p112(t):** Martin Hughes-Jones/Alamy Stock Photo; **p112(b):**
Sukree Sukplang/REUTERS; **p115:** Shutterstock; **p116:** © Crown copyright/
Contains public sector information licensed under the Open Government
Licence v3.0; **p119:** Iain McGillivray/Alamy Stock Photo; **p120:** Lawrence
Berkeley National Lab/Roy Kaltschmidt; **p123:** Ella Ling & Malaria No
More UK; **p126(t):** Global Initiative for Asthma; **p127(b):** Shutterstock;
p130: www.worldmapper.org; **p131:** Shutterstock; **p134:** TMAX/Fotolia;
p135(t): Shutterstock; **p135(b):** Used with kind permission of Michael
Crisafulli/www.VernianEra.com; **p138(t):** Courtesy of GRID Arendal www.
grida.no/graphicslib/detail/number-of-extra-skin-cancer-cases-related-to-uv-
radiation_1456; **p138(b):** EcoHealth, Climate Change and Global Health:
Quantifying a Growing Ethical Crisis, Volume 4, 2007, p397-205, Jonathan
A. Patz, Holly K. Gibbs, Jonathan A. Foley, Jamesine V. Rogers, Kirk R. Smith,
With permission of Springer; **p142:** ATTA KENARE/Getty Images; **p143:** OS
data © Crown copyright and database right 2018; **p144, 148:** Tim Bayliss;
p149: Eye Ubiquitous/Alamy Stock Photo; **p150:** Tim Bayliss; **p153, 163:**
Shutterstock; **p168:** Gerd Ludwig/National Geographic Creative; **p171:**
Shutterstock; **p172:** Rodrigo Baleia/Greenpeace; **p174, 177:** Shutterstock;
p177: Shutterstock; **p178:** David McNew/Getty Images.

Artwork by Kamae Design, Aptara Inc., and Ian West (**p68**).

Every effort has been made to contact copyright holders of material
reproduced in this book. Any omissions will be rectified in subsequent
printings if notice is given to the publisher.

Contents

Contents

Contents

We've got it covered, so you've got it covered

Success at AS and A Level is a very personal and rewarding challenge, but within your grasp providing you:

- put the time in
- believe in yourself and give it your full and best effort
- embrace the wealth of advice and opportunities your teachers and this course have given you.

Your revision guide

This revision guide contains key revision points that you need to prepare for Paper 2 of the AQA GCE Geography specification. It meets both the content requirements of the A Level course (Human geography), but can equally well be used for the separate AS course (Human geography and geography fieldwork investigation).

Both this and its sister *Physical Geography Revision Guide* need to be used alongside Oxford's AQA Human and Physical Geography Student Books. The revision guide provides links to the student books where necessary.

You will also find helpful other resources in this series including Oxford's *AQA Geography A Level and AS Exam Practice and Skills* book and *Kerboodle* digital resources and assessment online.

This revision guide provides:

- targeted AS and A Level revision guidance including top tips for exam success
- easy to digest spec-specific content, recaps and summaries
- flexible revision checklists to help you monitor your progress.

Key content of each section in the student book is summarised in a double or single page (Figure **1**).

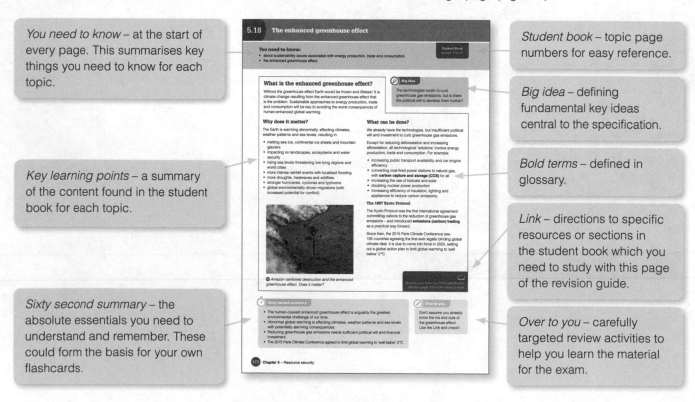

You need to know – at the start of every page. This summarises key things you need to know for each topic.

Key learning points – a summary of the content found in the student book for each topic.

Sixty second summary – the absolute essentials you need to understand and remember. These could form the basis for your own flashcards.

Student book – topic page numbers for easy reference.

Big idea – defining fundamental key ideas central to the specification.

Bold terms – defined in glossary.

Link – directions to specific resources or sections in the student book which you need to study with this page of the revision guide.

Over to you – carefully targeted review activities to help you learn the material for the exam.

⬥ **Figure 1** *Your revision guide's key features*

Guided answers to the activities can be found at www.oxfordsecondary.co.uk/geography-answers

In addition, each chapter introduction contains information on how the content relates to both AS and A Level specifications. You are also encouraged to build a record of essential key terms, and **either** track your revision progress **or** use the guidelines to indicate topics you are more or less confident about.

What does the AS specification include?

The AS Geography qualification does not count towards an A Level – it is a qualification in its own right.

The specification consists of six topics, of which you must study three plus a fieldwork investigation. They are grouped into two papers: *Physical geography and people and the environment; Human geography and geography fieldwork investigation.*

Paper 1: Physical geography and people and the environment	Paper 2: Human geography and geography fieldwork investigation
• Section A: you must study **one** of *Water and carbon cycles* **or** *Coastal systems and landscapes* **or** *Glacial systems and landscapes.* • Section B: you must study **either** *Hazards* **or** *Contemporary urban environments.*	• Section A: *Changing places.* **This topic is compulsory**. • Section B: *Geography fieldwork investigation and geographical skills.* You will have to answer one question on each of these topics.

Remember, there is no coursework at AS Level, but questions will be included in Paper 2 about the fieldwork you have completed as part of your AS course.

What does the A Level specification include?

The A Level specification consists of eleven topics of which you must study six plus a fieldwork investigation. They are grouped into three papers: *Physical geography; Human geography; Geography fieldwork investigation.*

Paper 1: Physical geography	Paper 2: Human geography	Paper 3: Geography fieldwork investigation
• Section A: *Water and carbon cycles.* **This topic is compulsory**. • Section B: you must study **one** of *Hot desert systems and landscapes* **or** *Coastal systems and landscapes* **or** *Glacial systems and landscapes.* • Section C: you must study **either** *Hazards* **or** *Ecosystems under stress.*	• Section A: *Global systems and global governance.* **This topic is compulsory**. • Section B: *Changing places.* **This topic is compulsory**. • Section C: you must study **either** *Contemporary urban environments* **or** *Population and the environment* **or** *Resource security.*	• You must complete an individual investigation which must include data collected in the field. • Your investigation must be based on a question or issue defined and developed by you relating to any part of the specification content.

Over to you

Complete this for Paper 2: Human geography

My optional topic for Section C is

....

.... (Chapter ....).

How to revise productively

Think about this introduction's opening comments on the challenge of AS and A Level. Revision doesn't have to be a bore or even a chore. With the right mindset it can be enormously satisfying, not least in the latter stages when all the pieces of the jigsaw start to come together, and you're anticipating what's coming next.

Rather like the 'solution' to predicting earthquakes, you and your friends' GCSE revision experiences should have convinced you that there is no 'one size fits all' solution to effective learning, memorising and/or revising. However, you'll know by now that revision has to be:

- scheduled, planned and **organised**
- **active** if it is going to be productive
- done **without distractions**
- in tune with your body clock and concentration patterns.

You are strongly encouraged to embrace the general 'good practice' explained by your teachers and the tips provided in this guide.

Use revision techniques that work for you, but don't avoid trying new ideas. You never stop 'learning to learn', and a new learning, memorising or revision aid tried now may prove to be invaluable – not just in these examinations, but for those at a higher level in years to come.

Think back to your GCSE revision. Reflect on how successfully (or not!) you:

- applied advice and guidance on planning your revision
- kept to and adapted your revision timetable if and when necessary
- experienced and understood your concentration curve (Figure **4**)
- managed your working environment, stress and distractions
- used different techniques
- practised using past exam questions.

Positive reflection, with the intention of learning from both your good and bad experiences of revision will pay dividends. So:

- consider the range of techniques you could use
- adopt what you know to be valuable
- be prepared to adapt and change.

I REALLY CRAMMED LAST NIGHT.

🔺 *Figure 2*

"It's called 'reading'. It's how people install new software into their brains"

🔺 *Figure 3*

 Over to you

Remind yourself of revision techniques, and add your own in the spaces provided. Tick the ones that work for you or add your own alternatives.

Mind-mapping (spider diagrams)	
Mnemonics	
Flashcards	
Topic posters	
Factfiles	
Lists of key terms/words	

Whatever 'old favourites' worked for you – use them again. Just because you were revising for GCSE doesn't devalue them at AS or A Level. But also consider taking your revision to the next level – there might be something here that transforms the way you work (Figure **5**).

Finally, you've probably already tried revising with a partner or in small groups. Message, Skype and Facetime interaction can be productive for some, but a persisting distraction if not scheduled and executed with real discipline.

Revising in class also needs self-discipline, but you should never underestimate the value of extra time spent with your teachers discussing topics you're not clear about, looking at exam questions and understanding mark schemes. However, it is revising alone that will dominate at this level – so make it count!

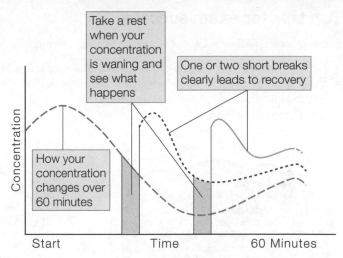

▲ **Figure 4** *How short rests boost concentration*

Fixed word count abstracts
Set yourself a fixed word count (say 50 or 100) for a topic summary. Once drafted, 'pruning' to the exact word count forces your mind to concentrate on identifying key information. Writing out the finished summary fixes knowledge into your long-term memory, so is more easily recalled when next revised.

Dealing with distractions
- Be an active learner – focus on the specific task, not the specified time.
- Leave your phone in another room – you can check it during your regular breaks.
- Divide each study session into short-range goals which demand your full attention.

Beware the highlighter pen!
Never highlight this guide or your notes straight away:
- 'Dribble' in soft erasable pencil in the early revision stages.
- Only use the ink highlighter when you're absolutely certain.
- Only use a colour you like.

Taking your revision to the next level

Revising key terms and command words
You'll have revision sessions when it just isn't going in, so change tack. For example:
- Focus on updating your revision progress and compiling the key terms in your chapter introductions (make use of the glossary).
- Remember – one of the biggest exam pitfalls is misunderstanding command words. So check and revise them.

Improving long-term memory by repetition
66% of material is lost within seven days if not reviewed again – and 88% is gone after six weeks! So build *review* time into your daily and weekly revision timetable. It will save you having to re-learn material from scratch!

Reading better and faster – the PQ2R method
P = Preview Begin with a quick skim to get an overview. Look for section headings, figure captions and key words.
Q = Question Look for answers to 'What?', 'Why?', 'Where?', 'When?' and 'Who?' questions which will identify the main learning point(s). This is active learning.
R = Read Now read the spread carefully, with these questions in mind. Your mind will actively look for answers. Make brief summary notes. Slow down over difficult sections.
R = Review Check your understanding by reviewing and testing your recall. Check your notes and answer the initial questions.

Practising output
Remember – exam preparation involves giving out information, not just taking it in! So:
- work on past exam papers
- follow the marks – see how marks are allocated by using the mark schemes.
- try a complete exam paper against the clock.

This will help you to avoid major surprises come the 'real thing'.

▲ **Figure 5** *Taking your revision to the next level*

Top tips for exam success

Success in your AS or A Level exam will involve three ingredients:

- having a thorough knowledge and understanding of the subject matter
- making the most of this knowledge and understanding in the exam through effective exam strategy and technique
- managing stress (Figure **6**).

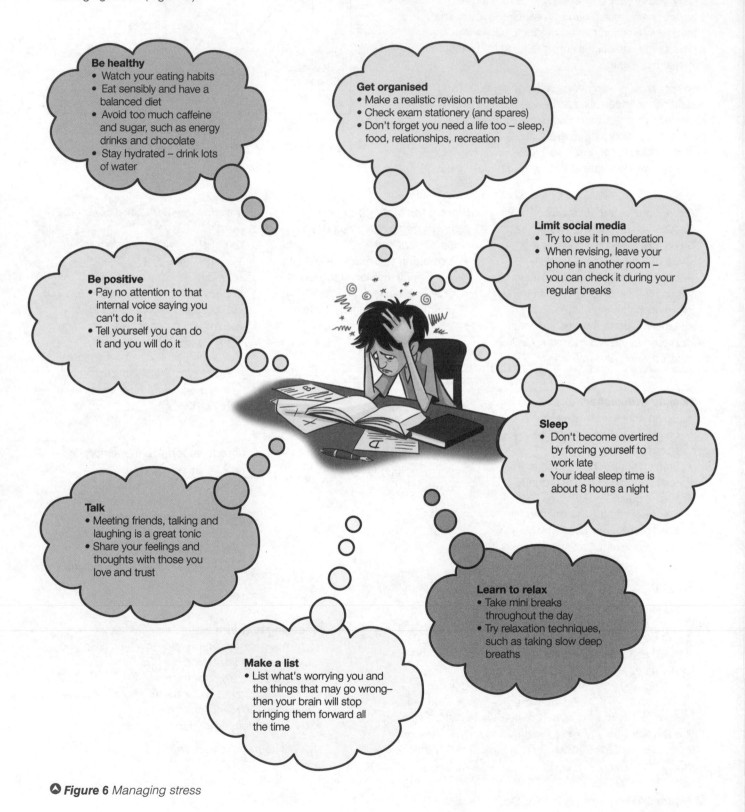

Be healthy
- Watch your eating habits
- Eat sensibly and have a balanced diet
- Avoid too much caffeine and sugar, such as energy drinks and chocolate
- Stay hydrated – drink lots of water

Get organised
- Make a realistic revision timetable
- Check exam stationery (and spares)
- Don't forget you need a life too – sleep, food, relationships, recreation

Limit social media
- Try to use it in moderation
- When revising, leave your phone in another room – you can check it during your regular breaks

Be positive
- Pay no attention to that internal voice saying you can't do it
- Tell yourself you can do it and you will do it

Sleep
- Don't become overtired by forcing yourself to work late
- Your ideal sleep time is about 8 hours a night

Talk
- Meeting friends, talking and laughing is a great tonic
- Share your feelings and thoughts with those you love and trust

Learn to relax
- Take mini breaks throughout the day
- Try relaxation techniques, such as taking slow deep breaths

Make a list
- List what's worrying you and the things that may go wrong– then your brain will stop bringing them forward all the time

⬢ **Figure 6** *Managing stress*

Your golden rules

Students who perform well almost always follow these rules.

They **revise** – thoroughly and productively (pages 8–9).

They know which **topics** will be in each exam (page 12).

They **practise** answers under timed conditions, including whole papers.

They are familiar with **mark schemes** and know what examiners are looking for to reach the highest **levels** (pages 13–14).

They **read the front cover instructions** – calming themselves by confirming what they expect.

They display and check an **exam timetable** with exact details of what, where and when (paper, venue, day and time).

They **read the paper carefully** – allowing time to 'size-up' and dissect questions before starting (Figure **7**).

Note the command – in effect 'define'

What **is meant** by the term 'global commons'? (4 marks)

The focus of the question – in this case, the term to be defined

This is a 4-mark question, so it must be point marked (correct points and development)

⬆ **Figure 7** Dissecting the question

They understand and respond correctly to **command** words (page 15).

They look at the **marks**, which indicate which questions will be point marked and which level marked (page 13 and Figure **7**).

They **answer every question required** and leave no blanks.

They write in **full sentences** using appropriate, **specialist vocabulary**.

They apply **specific located details** about case studies or examples.

They work out in advance and practise how long to spend on questions.

They never 'steal' extra time on favourite topics to the detriment of others.

They get the **timing** right – especially on longer answer questions – including 5–10% planning time, 80–85% writing time and 10% for a final read through to **check** and tidy-up errors.

How will you be assessed?
AS Level

There are two exams for AS Level Geography. Unlike A Level there is no coursework – but you must complete two days of AS fieldwork. The AS exams will involve:

Paper 1: Physical geography and people and the environment (1 hour 30 minutes)	Paper 2: Human geography and geography fieldwork investigation (1 hour 30 minutes)
There are a total of 80 marks available, worth 50% of the AS qualification. The paper consists of multiple-choice, short-answer levels of response and includes 9- and 20-mark extended prose questions. • Section A: **one** of *Water and carbon cycles* **or** *Coastal systems and landscapes* **or** *Glacial systems and landscapes* – **40 marks** • Section B: **either** *Hazards* **or** *Contemporary urban environments* – **40 marks**	There are a total of 80 marks available, worth 50% of the AS qualification. The paper consists of multiple-choice, short-answer levels of response and includes 9- and 20-mark extended prose questions. • Section A: *Changing places* – **40 marks** • Section B: *Fieldwork skills* – **40 marks**

A Level

There are two exams for A Level Geography and one individual fieldwork investigation – you must complete at least four days of A Level fieldwork. The A Level assessment will involve:

Paper 1: Physical geography 120 marks (40%)	Paper 2: Human geography 120 marks (40%)	Paper 3: Geography fieldwork investigation 60 marks (20%)
The 2 hours 30 minutes paper consists of multiple-choice, short-answer levels of response and includes 20-mark extended prose questions. • Section A: *Water and carbon cycles* – **36 marks** • Section B: **one** of *Hot desert systems and landscapes* **or** *Coastal systems and landscapes* **or** *Glacial systems and landscapes* – **36 marks** • Section C: **either** *Hazards* **or** *Ecosystems under stress* – **48 marks**	The 2 hours 30 minutes paper consists of multiple-choice, short-answer levels of response and includes 20-mark extended prose questions. • Section A: *Global systems and global governance* – **36 marks** • Section B: *Changing places* – **36 marks** • Section C: **either** *Contemporary urban environments* **or** *Population and the environment* **or** *Resource security* – **48 marks**	You must complete an individual investigation based on a question or issue defined and developed by you relating to any part of the specification content. You are advised to write between 3000 and 4000 words. Your investigation will be marked by your teachers and moderated by AQA.

How will your exam papers be marked?

Examiners have to know what it is that they are assessing you on, so they use Assessment Objectives, or AOs for short. There are three AOs for AS and A Level.

- **AO1: Demonstrate knowledge and understanding** of places, environments, concepts, processes, interactions and change, at a variety of scales (30–40%).
- **AO2: Apply knowledge and understanding** in different contexts to interpret, analyse and evaluate geographical information and issues (30–40%).
- **AO3: Use a variety of relevant quantitative, qualitative and fieldwork skills** to:
  o investigate geographical questions and issues
  o interpret, analyse and evaluate data and evidence
  o construct arguments and draw conclusions (20–30%).

Oxford's *AQA Geography A Level and AS Exam Practice and Skills* book will help you make sense of all this. It includes essential guidance on effective examination techniques, such as planning and structuring your longer answers and essays, e.g. **BUD**ing (**B**ox the command, **U**nderline the key words, **D**issect the question) and **PEEL**ing. Furthermore, at A Level it will help you to:

- understand the geographical themes/concepts that run across the course (Figure **9**)
- think about the course as a whole and how to approach it synoptically
- respond to questions about links between topics (or between sub-topics) more effectively.

Understanding the mark schemes

There are two types of mark scheme: point marked (up to 4 marks); and level marked (for 6 marks or more).

Point marking – short answers (worth up to 4 marks)

Questions carrying up to 4 marks are **point marked**. For every correct point that you make, you earn a mark. Sometimes these are single marks for a 1-mark question. Others require the development of points for additional marks. For example, if you are asked to describe one feature of something for 3 marks, development will be required.

> **What is meant** by the term 'global commons'? **(4 marks)**

There are four marks for this question and you have to define (or express what you understand by) the term 'global commons'. You would receive one mark per valid point, for example:

global commons are those parts of the planet that fall outside national jurisdictions (√) and to which all nations have access (√)

with additional credit for development and/or exemplification, for example:

four are identified (√) – the high seas and deep oceans, the atmosphere, the northern and southern polar regions, and outer space (√).

Level marking – extended answers (worth 6 marks or more)

Questions carrying 6 marks or more are **level marked**. The examiner reads your whole answer and then uses a set of criteria – known as levels – to judge its qualities. There are two levels for questions carrying 6 marks, three levels for 9 marks, and four levels for 20 marks.

Think of levels in a mark scheme as a staircase you have climb in order to reach the top. Each step up to the next level, demands a little bit more of you (Figure **8**).

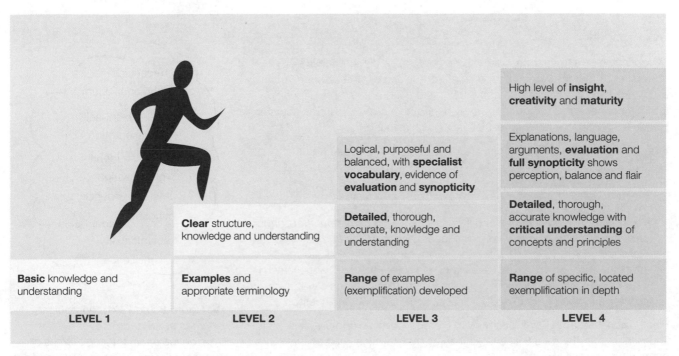

LEVEL 1	LEVEL 2	LEVEL 3	LEVEL 4
Basic knowledge and understanding	**Examples** and appropriate terminology	**Range** of examples (exemplification) developed	**Range** of specific, located exemplification in depth
	Clear structure, knowledge and understanding	**Detailed**, thorough, accurate, knowledge and understanding	**Detailed**, thorough, accurate knowledge with **critical understanding** of concepts and principles
		Logical, purposeful and balanced, with **specialist vocabulary**, evidence of **evaluation** and **synopticity**	Explanations, language, arguments, **evaluation** and **full synopticity** shows perception, balance and flair
			High level of **insight**, **creativity** and **maturity**

⬆ **Figure 8** *Example steps in level marking*

Introduction: a guide to success

Figure **8** shows general principles of level marking, but the actual criteria – determined by the AOs – will vary according to the nature of the question set. Hence the importance of examining and discussing mark schemes with your teachers.

For example, 9-mark questions have three levels in their mark schemes. The balance of marks is split between 4 marks for AO1 and 5 marks for AO2. So to achieve Level 3 you should aim to:

- demonstrate accurate knowledge and understanding throughout (AO1)
- apply your knowledge and understanding (AO2)
- produce a full interpretation that is relevant and supported by evidence (AO2)
- make supported judgements in a balanced and coherent argument (AO2).

Mark schemes for 20-mark questions have four levels. The balance of marks is split evenly with 10 marks for AO1 and 10 marks for AO2. So to achieve Level 4, in addition to the qualities for Level 3 you should aim to:

- reach a detailed evaluative conclusion that is rational and firmly based on knowledge and understanding which is applied to the context of the question (AO2)
- show a detailed, coherent and relevant analysis and evaluation in the application of knowledge and understanding throughout (AO2)
- show full evidence of links between knowledge and understanding and the application of knowledge and understanding in different contexts (AO2).

Examples and case studies

Case studies are in-depth examples of particular places, used to illustrate big ideas at localised scales. Unlike GCSE, few questions at AS and A Level specifically ask for examples, but you can only access higher levels of the mark schemes by including them to help you better demonstrate the points you make.

The important thing is that discussion of located, named examples makes your answer precise.

But be selective – no question at this level will ever ask you to write everything you know about a case study.

For the *Changing Places* topic you have to undertake two in-depth *place studies*: one local, the other contrasting and distant. These are *localities* (small-scale places) that you will be asked to write about in Paper 2, so ensure you can write confidently about the developing character of both of these places (see 2.9 and 2.12).

Synopticity

Synopticity refers to your capability to demonstrate a comprehensive understanding of the whole picture. In effect it is demonstrated when your exam answers show evidence of the complete geographer – the capacity to draw connections and supporting evidence from *anywhere* in the course. Hence its inclusion in the highest levels of level marking.

Indeed, the higher tariff questions are likely to ask you to make **links**. For example:

- **links** between subtopics within a compulsory topic, such as *Changing Places*
- **links** between elements of two different topics such as *factors that shape the character of a place* (Changing Places) and *the form and nature of interdependence* (Global Systems).
- **links** between topics that you have learnt and novel situations or phenomena that are not in the specification; these type of questions will come with a resource and you'll have to apply what you know to answer the question.

Finally, don't fret about synopticity. With thorough revision and practice you will get there! Figure **9** illustrates the key geographical themes that run across the course.

thresholds
causality equilibrium
systems feedback
representation
inequality identity
globalisation
interdependence
mitigation adaptation
sustainability
resilience risk

⬆ *Figure 9 The key geographical themes at AS and A Level. Can you define each of these themes or concepts? Can you give real world examples for more than one topic?*

Introduction: a guide to success

Exam question command words

These quite simply tell you what to do. You cannot answer questions properly if you don't understand them – so when you first read a question, check out the command word. It is *crucial* to ensure that you recognise and understand these commands instinctively before you sit your examinations.

Command word	What is it asking you to do?	Assessment objective
Analyse	Break down concepts, information and/or issues to convey an understanding of them by finding connections and causes and/or effects.	AO2 and AO3
Annotate	Add to a diagram, image or graphic a number of words that describe and/or explain features, rather than just identify them (which is labelling).	AO3
Assess	Consider several options or arguments and weigh them up so as to come to a conclusion about their effectiveness or validity.	AO1 – but mainly AO2
Compare	Describe the similarities and differences of at least two phenomena.	AO1 or AO3
Contrast	Point out the differences between at least two phenomena.	AO1 or AO3
Critically	Often occurs before 'Assess' or 'Evaluate' inviting an examination of an issue from the point of view of a critic with a particular focus on the strengths and weaknesses of the points of view being expressed.	AO1 – but mainly AO2
Define..., What is meant by...	State the precise meaning of an idea or concept.	AO1
Describe	Give an account in words of a phenomenon which may be an entity, an event, a feature, a pattern, a distribution or a process. For example, if describing a landform say what it looks like, give some indication of size or scale, what it is made of, and where it is in relation to something else (field relationship).	AO1
Distinguish between	Give the meaning of two (or more) phenomena and make it clear how they are different from each other.	AO3
Evaluate	Consider several options, ideas or arguments and form a view based on evidence about their importance/validity/merit/utility.	AO1 – but mainly AO2
Examine	Consider carefully and provide a detailed account of the indicated topic.	AO1
Explain... Why... Suggest reasons for...	Set out the causes of a phenomenon and/or the factors which influence its form/nature. This usually requires an understanding of processes.	AO1 and AO2
Interpret	Ascribe meaning to geographical information and issues.	AO3
Justify	Give reasons for the validity of a view or idea or why some action should be undertaken. This might reasonably involve discussing and discounting alternative views or actions.	AO1 – but mainly AO2
Outline..., Summarise...	Provide a brief account of relevant information.	AO1
To what extent...	Form and express a view as to the merit or validity of a view or statement after examining the evidence available and/or different sides of an argument.	AO1 – but mainly AO2

1 Global systems and global governance

(AL) *Global systems and global governance* is a **core topic**. You must answer **all** questions in Section A of Paper 2: Human geography.

Paper 2 carries 120 marks and makes up 40% of your A Level. Section A carries 36 marks.

Specification subject content (specification reference in brackets)

Either tick these boxes as a record of your revision, or use them to identify your strengths and weaknesses

Section in Student Book and Revision Guide	1 ☹	2 😐	3 🙂	Key terms you need to understand Complete the key terms (not just the words in bold) as your revision progresses. 1.1 has been started for you.
Globalisation *(3.2.1.1)*				
1.1 What is globalisation?				*transnational companies, flows, factors, less developed economies (LDEs), highly developed economies (HDEs)*
Global systems *(3.2.1.2)*				
1.2 Interdependence and unequal flows				
1.3 The internet and single-product economies				
International trade and access to markets *(3.2.1.3)*				
1.4 International trade				
1.5 Trading relationships				
1.6 Trade agreements and access to markets				
1.7 Transnational corporations				

1.8 World trade: Coca-Cola			
1.9 World trade: fair trade?			
1.10 Global food systems			
Global governance *(3.2.1.4)*			
1.11 Global governance			
1.12 Global governance: issues and inequalities			
The 'global commons' *(3.2.1.5)*			
1.13 The global commons: what is it?			
Antarctica as a global common *(3.2.1.5.1)*			
1.14 Antarctica: threats from fishing, whaling and mineral exploitation			
1.15 Antarctica: scientific research and climate change			
1.16 Antarctica: tourism			
1.17 Antarctica: global governance			
Globalisation critique *(3.2.1.6)*			
1.18 Globalisation critique			

Note: Specification reference **3.2.1.7** refers to skills

You need to know:
- dimensions of globalisation
- factors in globalisation.

Student Book
pages 8–9

What is globalisation?

Globalisation is a *multi-faceted* process of change that has led to an increasingly interconnected world, in particular, over recent decades. Globalisation:

- includes opening-up world trade and markets to **transnational corporations (companies) (TNCs)**
- results in a wide a range of associated effects on people, culture, political systems and the environment.

Dimensions of globalisation

Many different, and often complex, *flows* connect the world today, including the flow of:

- information – e.g. shared worldwide within and between TNCs
- capital – *highly developed economies (HDEs)* invest in *less developed economies (LDEs)* to take advantage of cheaper production costs
- technology – the internet largely ignores political boundaries
- products – cheap global transport networks have encouraged (intercontinental) trade in products
- labour – efficient transport enable workers and tourists to travel increasing distances between places
- services – follow the flows of capital, information, people and products.

TNCs produce **global products** with the same global brands the world over, as part of their *global marketing* strategies (see 1.7).

- Products, often manufactured by TNCs in lower cost LDEs, are subsequently widely distributed.
- Whilst populations of HDEs still consume *a* majority of global products, *a* growing share of the global consumer class now lives in EMEs, like China and India.

Factors influencing globalisation

- New technologies, communications and information systems – information may be shared easily and cheaply.
- Global financial systems – allow flows and lending of money but may also cause negative feedback/limits within the system (e.g. global banking crisis of 2008).
- Transport systems – the friction of distance (time and space) has declined, creating new opportunities but also threats (e.g. spread of disease).
- Security – cybersecurity is a growing concern for governments and companies.
- Trade agreements – the **World Trade Organisation (WTO)** ensures trade agreements between countries and trade blocs are followed.

 Big idea

Globalisation refers to an ever increasing number and complexity of connections between people and places, worldwide.

Figure 1 *The multi-faceted process of globalisation*

 Sixty second summary

- Globalisation is linked with opening up world trade and markets to TNCs.
- Impacts of globalisation affect every person on the planet.
- Globalisation has many interconnected aspects and dimensions.
- Several factors have influenced globalisation over recent decades, including the development of new technology, systems and relationships that relate to communications, finance, transport, trade agreements and security.

 Over to you

Sometimes it helps to sketch a complex idea on paper. Draw a **nine** point star and add **nine** key components (dimensions and factors) of globalisation.

You need to know:
- issues associated with unequal flows of people, money, ideas and technology.

Student Book
pages 10–11

Uganda and global systems

The increasing interdependence of economic, political, environmental and social global systems results in winners and losers.

The example of Uganda illustrates both positive and negative outcomes which result from unequal flows of people, money, ideas and technology.

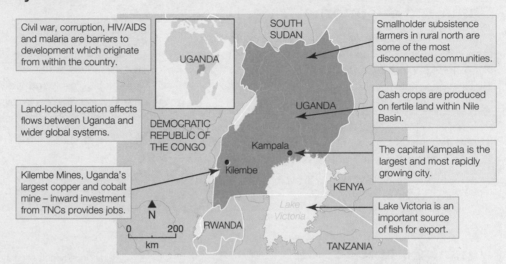

Civil war, corruption, HIV/AIDS and malaria are barriers to development which originate from within the country.

Land-locked location affects flows between Uganda and wider global systems.

Kilembe Mines, Uganda's largest copper and cobalt mine – inward investment from TNCs provides jobs.

Smallholder subsistence farmers in rural north are some of the most disconnected communities.

Cash crops are produced on fertile land within Nile Basin.

The capital Kampala is the largest and most rapidly growing city.

Lake Victoria is an important source of fish for export.

▲ **Figure 1** *Uganda is an LDE shaped by the form and nature of global systems*

Inequalities and injustices

Poverty and unequal flows of people exist across Uganda:

- Migration into and within Uganda has been driven by political factors, poverty and rapid population growth (there are few controls on Uganda's borders).
- Northern (rural) Uganda has faced economic, demographic and security challenges – due to civil war and also migrants from South Sudan.
- TNC investment is greatest around the capital city, Kampala, and where resources such as cobalt, gold, copper and iron ore are extracted.
- TNCs offer much needed foreign investment, but are criticised for working within a culture of corruption, bribery and mismanagement. Low value cash crops (coffee, tea, cotton) once traded by the British East Africa Company, still dominate exports nationally.
- Overfishing, and predatory Nile Perch (introduced by the British) have resulted in the near extinction of local fish stocks.

Stability growth and development

Shambas, the smallholdings owned by subsistence farmers, show the positive effects of global systems.

Mobile phone antennae are evidence of recent technological changes.

- Cheap wireless technology has enabled telephony to be introduced into these remote rural areas.
- The 'Village Phone' model offers loans to business people wishing to start a mobile phone business.
- Each loan allows purchase of a mobile smart phone, car battery to charge it and booster antenna to pick up signals from 25 km away.
- Farmers pay to use the service to access market prices, weather data, farming techniques and so on.

Sixty second summary

- Uganda's economy is shaped by unequal flows of people, money, ideas and technology, within and across its borders, as part of wider global systems.
- Poverty in Uganda affects one-third of the population and is still greatest in rural areas.
- The Village Phone model is one example of the way in which new communications technology has helped to promote economic development in poor rural settlements.

Over to you

Two column tables are excellent tools for summarising both sides of an argument. This helps to ensure a balanced approach to an answer. Using a simple table, list the pros and cons of globalisation for Uganda's economy and population.

You need to know:
- how unequal power relations and use of the internet influence geopolitical events
- the negative impacts of a single-product economy, within the wider context of global economic systems.

Student Book
pages 12–13

China and the internet

The internet connects more than 4 billion people around the world. It illustrates many of the benefits and challenges of globalisation, by enabling flows of money, ideas and technology.

- China has the largest number of internet users in the world.
- Alibaba.com is a highly successful Chinese e-commerce company which enables domestic consumers to buy branded products from international as well as Chinese companies.

- The 'Great Firewall' is a system of online censorship operated by the Chinese central government which blocks access to foreign websites and also slows internet service (**bandwidth throttling**).
- 'The Golden Shield' is a system of domestic surveillance that further helps to enforce state-led censorship.

Nigeria's single-product economy

Nigeria is a country that should have done well from globalisation. It is an example of a **single-product economy**.

- Oil and gas revenue accounts for more than 80% of export revenues.
- Membership of **OPEC** has aided Nigeria's economic growth.
- Focus on oil and gas alone has resulted in dramatic decline in traditional industries of agriculture and manufacturing.
- Oil and gas TNCs have been encouraged to exploit the reserves despite environmental concerns and disregard of local needs, including land rights.
- Effects of **Dutch disease** – high incomes generated from oil/gas sales mean imported consumer goods are cheaper than domestically manufactured goods. This leads to deindustrialisation.

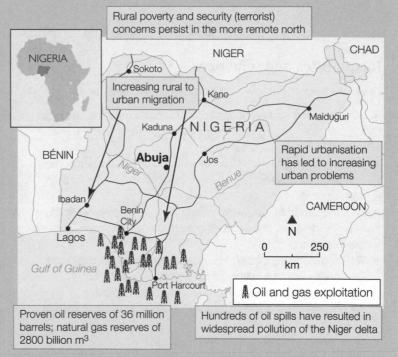

Rural poverty and security (terrorist) concerns persist in the more remote north

Increasing rural to urban migration

Rapid urbanisation has led to increasing urban problems

 Oil and gas exploitation

Proven oil reserves of 36 million barrels; natural gas reserves of 2800 billion m³

Hundreds of oil spills have resulted in widespread pollution of the Niger delta

▲ **Figure 1** *Nigeria – an economy dominated by oil and gas exports*

Sixty second summary

- Some countries use global systems to their own advantage – such as China's control of the internet.
- China uses the 'Great Firewall' in addition to more traditional methods of state-controlled censorship to control and influence geopolitics.
- Global trade links are exploited by China (manufactured goods) and Nigeria (oil and natural gas).
- Nigeria's oil and gas exports provide much needed foreign income, but has resulted in expensive domestic goods and a reliance on foreign-owned TNCs.

Over to you

Consideration of future events tends to be a feature of higher-order responses. Suggest how the economy of Nigeria may change in your lifetime; think about how these changes might impact Nigeria's development.

Student Book
pages 14–17

You need to know:

• global trends and patterns in the volume and pattern of international trade and investment.

Trends in international trade

International trade is:

• *inevitable* – no country has all the materials and resources that it requires
• *contentious* – foreign products may be bought in cheaply but the domestic seller may lose a sale
• *political* – countries exert power to ensure national gains
• the *key driver* of global economic growth – it has increased dramatically, particularly since the Second World War
• not always *free trade* – LDEs continue to struggle to trade within trading blocs of HDEs.

Advantages	Disadvantages
• *Comparative advantage* – specialisation produces goods at lowest cost	• *Over-specialisation* – leads to potential vulnerability if demand for goods declines
• *Economies of scale* – cheaper cost per unit	• *Product dumping* – exporting at lower price than price charged domestically
• *Purchasing power* – increased competition lowers prices	• *Slow growth or decline of local industries* – a result of foreign competition
• *Fewer domestic monopolies* – as a result of (cheaper) imports	• *Protectionist reactions* – when government imposes additional taxes on imported goods
• *Transfer of technology* – use of new technologies to reduce costs	• *De-skilling* – traditional skills lost to production technology and 'job flight'; so in HDEs, 'screwdriver jobs' dominate
• *Increased employment* – as result of production for export	• *New international division of labour* is exploitative – workers' rights compromised in LDEs, as labour costs are squeezed by TNCs that subcontract on a short-term basis (see 1.7).

⬆ *Figure 1 Advantages and disadvantages of international trade*

Songxia – Umbrella City

Around 70% of the world's umbrellas are made in China. Around 500 000 are made annually in Songxia, the 'umbrella capital of the world'.

Good road network to population centres (e.g. Hangzou City, Shanghai) and ports (e.g. Ningbo) for export.

Local government financial support, e.g. Songxia Umbrella Industrial Park.

1200 factories – a single worker makes 300 umbrellas a day.

Low labour costs – local female, flexible.

[Map of China showing: Harbin, Beijing, CHINA, Chongqing, Hangzhou, Shanghai, Songxia, Ningbo, Guangzhou]

⬆ *Figure 2 Comparative advantages of Songxia for umbrella production*

Continued over ▶▶▶

Patterns of international trade and investment

The value of exports have traditionally been dominated by North America, Europe and East Asia, although this pattern is beginning to change (Figure **3**).

- The poorest 50 countries account for less than 1% of global trade.
- The top five exporting countries account for over one-third of global trade.
- LDEs exports generally still dominated by limited number of low value **primary products** (and single-product economies).
- HDEs exports still tend to be dominated by consumer goods, although there is increasing competition from EMEs and the BRICS (Brazil, Russia, India, China, South Africa).
- **Foreign Direct Investment (FDI)** has tended to follow the patterns in world trade.
- HDEs still receive the majority global share of FDI but an increasing proportion is flowing into EMEs, particularly those in Asia.

	1948	1953	1963	1973	1983	1993	2003	2013
Value (US$ billion)								
World	59	84	157	579	1838	3684	7380	18301
Share (%)								
World	100.0	100.0	100.0	100.0	100.0	100.0	100.0	100.0
North America	28.1	24.8	19.9	17.3	16.8	18.0	15.8	13.2
South and Central America	11.3	9.7	6.4	4.3	4.5	3.0	3.0	4.0
Europe	35.1	39.4	47.8	50.9	43.5	45.3	45.9	36.3
Commonwealth of Independant States (CIS)	–	–	–	–	–	1.5	2.6	4.3
Africa	7.3	6.5	5.7	4.8	4.5	2.5	2.4	3.3
Asia & Oceania	14.0	13.4	12.5	14.9	19.1	26.0	26.1	31.5

Value of international trade has grown dramatically since end of Second World War.

Export of manufactured goods is traditionally dominated by North America.

Decline of HDEs share over last 30 years. Increased share of BRICS (Brazil, Russia, India, China, South Africa).

China overtook Japan in 2004 and USA in 2007 to become the leading exporter.

 Figure 3 *World merchandise exports by region, 1948–2013*

Sixty second summary

- Value and volume of trade has increased significantly since the Second World War.
- There are positive and negative impacts of international trade – although arguably, HDEs tend to benefit more than LDEs.
- Songxia, Umbrella City, is an example of specialised production.
- Patterns of international trade are changing – EMEs' and LDEs' market share of exports is increasing.
- Small but significant shift in share of world trade from HDEs to developing economies and the BRICS countries.
- China is the largest recipient of foreign direct investment.

Over to you

Summarise trends and patterns of international trade and investment in **three** bullet points. **Each** point should be supported by an appropriate number. (Use data selectively in support of an answer and round up or down to make comparison easier. Don't forget you can use a calculator in the exam).

Student Book
pages 18–19

You need to know:

- about terms of trade
- the impacts of trade in metals.

Terms of trade

Terms of trade refers to cost of goods that a country has to import compared with price at which they can sell the goods they export.

- Terms of trade traditionally favour HDEs, as cheaper primary products are imported from LDEs and then manufactured into higher value (consumer) goods by HDEs (Figure **1**).
- Because there is a widening price gap between value of primary products and value of manufactured goods, some LDEs need to export increased volumes of primary products if they are to afford to buy manufactured goods from HDEs.

Coffee

Growers 10%

Exporters 10%

Shippers and roasters 55%

Retailers 25%

LDE and EMEs export primary products, such as coffee, to HDEs for manufacture.

Huge value added to product after the manufacturing process, usually in HDEs. Retailers, particularly TNCs, selling global products benefit from a large share of the profits.

⬥ **Figure 1** *Terms of international trade frequently favour HDEs.*

Sixty second summary

- All countries attempt to manipulate the terms of trade, or the price of exports compared to costs of imports, to their own advantage.
- Terms of trade tend to have a negative impact on LDEs.
- Demand for metal has increased, particularly in the EMEs of Asia.
- Overproduction and increased competition (from LDEs and EMEs) have resulted in declining steel production in HDEs.

Over to you

In 60 seconds, summarise changes in global trading relationships over the last 30 years.

Big idea

A global shift in trading relationships is taking place, with the increasing significance of exports from, and patterns of consumption of, EMEs. This represents a geographical and economic shift away from the HDE economies of the northern hemisphere.

Impacts of trade in metals

There has been increased demand for metal from the emerging economies of east Asia, particularly China (Figure **2**).

- Over one-third of all copper is consumed in China.
- To meet increased demands China has invested heavily in African mining operations (Figure **3**).
- New mines in Africa have caused a positive multiplier effect – including jobs and (government) investment.
- Critics argue that China is exploiting weaknesses in African economies and quote unsafe and unethical working practices.

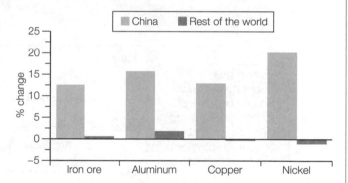

⬥ **Figure 2** *Percentage growth in China's consumption of metals (2002–14).*

Collum Coal Mining Industries, Zambia	Tata Steel, UK
Supplies fuel to Zambia's copper and cobalt mines.	Indian-owned (but merging with German steel manufacturer Thyssenkrupp).
11 African workers shot in 2011 by Chinese managers during protests against poor working conditions.	Chinese 'steel dumping' (where steel is sold at a rate below the cost to produce it) forces Tata Steel to announce job cuts across UK sites in 2016.
Following a 'poor safety, health and environmental record' and accusations of tax avoidance, the mine is seized by the Zambian Government in 2013.	UK Government avoids direct conflict with China as keen to encourage Chinese inward investment. Critics push for increased protectionism from cheap imports.

⬥ **Figure 3** *Impacts of trade in metals in Zambia and the UK*

You need to know:
- about trade blocs
- about trading agreements and their impact on economies and societies
- the role of the EU in trade agreements and governance.

Student Book
pages 20–3

Major trade blocs

A **trade bloc** is a group of countries, largely organised by geographical region, which supports free trade between member states. A common tariff may be paid on goods imported into the trading bloc from outside.

 Big idea

Virtually all international trade moves through trading blocs – they are hugely influential on patterns of international trade.

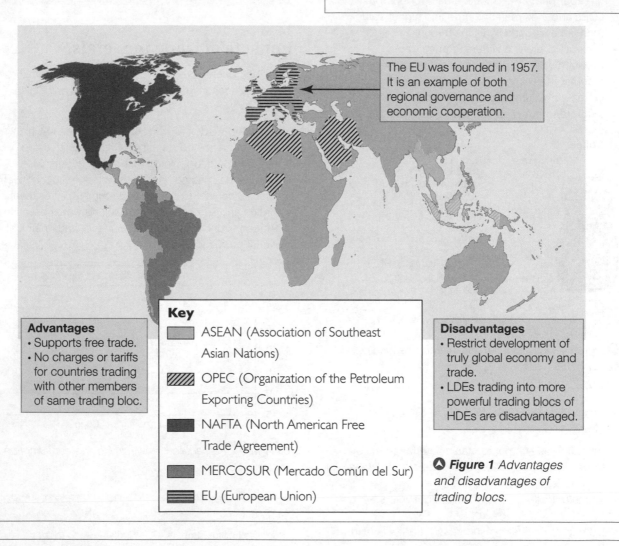

The EU was founded in 1957. It is an example of both regional governance and economic cooperation.

Advantages
- Supports free trade.
- No charges or tariffs for countries trading with other members of same trading bloc.

Key
- ASEAN (Association of Southeast Asian Nations)
- OPEC (Organization of the Petroleum Exporting Countries)
- NAFTA (North American Free Trade Agreement)
- MERCOSUR (Mercado Común del Sur)
- EU (European Union)

Disadvantages
- Restrict development of truly global economy and trade.
- LDEs trading into more powerful trading blocs of HDEs are disadvantaged.

⬥ **Figure 1** *Advantages and disadvantages of trading blocs.*

Trading organisations

Several organisations help to encourage trade by governing and setting rules:

- *World Trade Organisation (WTO)* – removes trade barriers such as tariffs.
- *Organisation for Economic Cooperation and Development (OECD)* – a global think tank.
- *Organisation of Petroleum Exporting Countries (OPEC)* – regulates global oil market.
- *World Bank* – promotes global investment and offers loans for (developing) countries.

- *International Monetary Fund (IMF)* – promotes global monetary and exchange stability.
- *G8* – the Group of 8 (Canada, France, Germany, Italy, Japan, Russia, UK, USA), the eight 'richest countries' of the world meet annually to discuss economic development (the G8+5 adds China, India, Brazil, Mexico, South Africa).
- *G20* – adds South Korea, Australia, Turkey, Saudi Arabia, Argentina, Indonesia and the leader of the EU and representatives of the IMF and World Bank.

Structure of the European Union

European Council
Sets the EU's overall political direction and priorities. It is made up of the heads of state or government of EU member countries.

European Commission
Responsible for proposing and implementing EU laws, monitoring treaties and the day-to-day running of the EU. Members are appointed by EU national governments.

European Parliament
Directly elected by the 500 million EU citizens. Adopts the laws proposed by the Commission. Shares power over EU budget and legislation with the Council of the European Union.

Council of the European Union
Represents the governments of member countries and promotes/defends national interests. Shares power over EU budget and legislation with the European Parliament.

Member countries
Implement the laws passed by the EU. The Commission ensures that the laws are properly applied and implemented.

🔺 **Figure 2** *The institutions of the EU have delivered over 60 years of stability, peace and prosperity for its member states.*

Sixty second summary

- Trading blocs and trading organisations link countries around the world.
- Member countries of a trading bloc may trade with each other without charges or tariffs (free trade).
- The EU is a large and influential trading bloc that has delivered peace and (relative) economic prosperity for 500 million people.
- There are arguments for and against membership of trade blocs such as the EU.
- Greece almost left the EU in 2015 (Grexit); in 2016 the UK voted to leave (Brexit).

Over to you

The importance of 'wider reading' can never be underestimated. Brexit will remain popular discussion points in the media for some time. Prepare and then subsequently talk for one minute on **either** the arguments for **or** against membership of the EU. Refer to at least **one** recent item of relevant news.

Greece and the European Union

Greece almost left the EU in 2015 (a so-called Grexit) because of huge debts to the EU. This was because of:

- unsustainable spending on the public sector
- massive government borrowing
- inadequate enforcement of taxation
- the global financial crisis of 2008.

In 2010 Greece ran out of money and had to borrow over 200 billion euros from the EU and the IMF. If Greece had retained its own currency, it could have been devalued to make its exports more competitive and its imports less so. As this was not possible with the euro, Greece was only able to become more competitive by shedding costs – one result of which was a sharp increase in unemployment, particularly within the public sector.

However, whilst cuts have undoubtedly hurt the Greek economy, its membership of the EU still offers the benefits of free trade.

Arguments FOR staying in the EU (against Grexit)

- Greece benefits from being part of the European free market (it imports nearly 50% of its food and 80% of its energy).
- Leaving the EU would change Greece's trade balance, might cause bankruptcies and high inflation with an associated social cost.
- Greece would find it very difficult to borrow further and would therefore have to pass on increased living costs to the population. Everything would be more expensive.

Arguments AGAINST staying in the EU (for Grexit)

- Greece would be able to trade more freely and to take advantage of its location and geography (e.g. might become a regional trading hub and gateway to the Middle East, Balkans or Russia).
- National needs might be met in a more sustainable and local way.
- The return of the drachma would see flexibility of exchange rates.
- Imports would be more expensive, but this may promote economic growth and create jobs.

You need to know:
- about the nature and role of TNCs
- the spatial organisation, linkages, production, impacts, trading and marketing patterns of TNCs.

Student Book
pages 24–7

What is a transnational corporation (TNC)?

TNCs are large, usually global, companies, which operate in more than one country. They:

- produce global products – these may be consumer goods that have a strong recognisable brand and which are distributed, marketed and sold in a large number of countries
- dominate all industrial sectors – primary (e.g. mining), secondary (e.g. food), tertiary (banking) and quaternary (pharmaceuticals).

Spatial organisation of TNCs and their model of production

TNCs usually adopt a hierarchical model of organisation, imposing top-down decision-making and control of activities. So branch plants (plants or factories whose HQ is in another country) tend to be recipients (rather than initiators) of change. This can take the form of either **vertical integration** (Figure **1**) or **horizontal integration** (Figure **2**).

Alternatively, TNCs may subcontract different manufacturing processes to other companies. This is known as *truncation* or *vertical disintegration* of production (Klein, 2000). This avoids costly pension schemes, as workers aren't directly employed by the TNC.

By working globally, TNCs are able to take advantage of:

- **spreading of locational risks** – the spatial organisation of TNCs minimises any impacts of regional events, e.g. natural hazards, political change, economic downturn
- **flexibility of production** – exploit regional differences in factors of production, e.g. costs of labour, raw materials, and land (Figure **3**).

One company controls or owns several stages in production and distribution chain. Increases control and power.

🔺 *Figure 1* Vertical organisation

Improves links between different firms in same stage of production. Usually occurs as result of acquisition of competitors in the same industry.

🔺 *Figure 2* Horizontal organisation

🔽 *Figure 3* BP's spatial organisation is an example of vertical integration

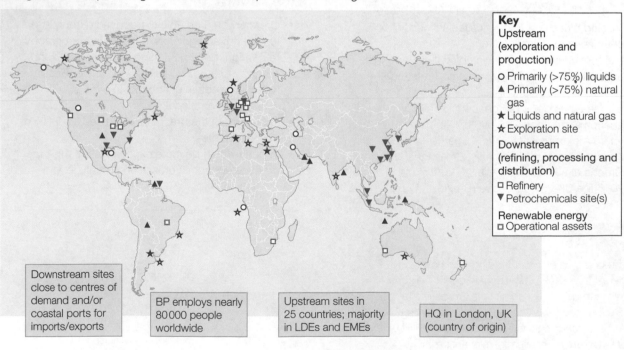

Key
Upstream (exploration and production)
- ○ Primarily (>75%) liquids
- ▲ Primarily (>75%) natural gas
- ★ Liquids and natural gas
- ✦ Exploration site

Downstream (refining, processing and distribution)
- ☐ Refinery
- ▼ Petrochemicals site(s)

Renewable energy
- ☐ Operational assets

Downstream sites close to centres of demand and/or coastal ports for imports/exports

BP employs nearly 80 000 people worldwide

Upstream sites in 25 countries; majority in LDEs and EMEs

HQ in London, UK (country of origin)

Reasons for growth of TNCs

- Cheap labour – usually from LDEs and EMEs but also from unemployed in HDEs.
- Mergers and takeovers – smaller competitors bought out; may lead to eventual **monopoly**.
- Flexible workforce – including working overseas.
- Availability of finance to fund expansion – includes investment by TNCs from EMEs.
- Fewer environmental restrictions – and possibly weakly enforced.

- Globalised transport network – including the impacts of the 'containerisation revolution'.
- Technological developments – such as refrigeration and freeze-drying of perishables.
- Government encouragement – tax incentives, subsidies and favourable policies.
- Cheap land – such as brownfield (former industrial) sites.

Impact of TNCs

	Favourable	Unfavourable
Host country	Increase employment and thereby raise living standards.	Many jobs are of low skill in LDEs.
	Improve levels of skill and expertise.	Managerial positions tend to be brought in rather than developed locally.
	Foreign currency brought in, improving the balance of payments.	Majority of profits are sent back to home country.
	Socio-economic multiplier effect, e.g. increased purchasing power, leads to demand for consumer goods and further economic growth.	Multiplier effects can also be negative, e.g. on the environment, concerns over safety and employee rights.
	Encourage a transfer of technology into the country, e.g. the growth of telecommunications.	Investment may only be short-term and TNC may pull out at short notice.
Country of origin	Development of higher-order jobs such as in research and development or management.	Workforce may need to relocate or make increased visits to operations overseas.
	Overseas investment adds to income for the whole nation (via tax and other multiplier effects).	As a result of loopholes, corporation (business) tax is not paid fully by all TNCs.
	Wider share ownership – individuals and companies more willing to become involved in foreign investments.	Speculative investments in TNCs for quick returns helped contribute to the global 2008 financial downturn.

Figure 4 *Impacts on countries in which TNCs operate*

Production, trading and marketing patterns of TNCs

The largest TNCs might have offices, branch plants and retail outlets in almost every country of the world. However:

- production generally takes place in LDEs or EMEs (particularly China)
- design, research and development tends to be in HDEs
- headquarters are in the country of origin
- they sell and market the same or similar product everywhere, by investment in 'building the brand'; although global products are tweaked to meet local needs (glocalisation).

 Sixty second summary

- TNCs operate in more than one country and produce global products.
- Most TNCs have top-down management within a hierarchical organisation.
- TNCs control the many stages of production via some form of horizontal or vertical integration.
- Since the Second World War, the number of TNCs has increased dramatically and with an increasingly complex spatial organisation.
- Many factors encourage the growth of TNCs, e.g. government incentives, the success of global marketing strategies.
- TNCs bring both positives and negatives to countries; profits frequently flow from centres of production to HDEs where most TNCs have their headquarters.

 Over to you

Write a clear definition of a transnational corporation (TNC). Write **three** positives and **three** negatives associated with the growth of TNCs.

You need to know:

* about impact of Coca-Cola on those countries in which it operates.

Student Book
pages 28–9

Coca-Cola – an example of a global product

Coca-Cola is one of the world's most recognised, and valuable brands.

* Nearly 2 billion Coca-Cola products are consumed daily.
* Growth has been the result of successful marketing, astute investment and strategic worldwide acquisitions.
* Whilst quick to realise new opportunities in expanding markets such as eastern Europe and India, Coca-Cola has been successful in spreading locational risks across nearly 200 countries.

Coca-Cola has had positive and negative impacts on host countries.

It is alleged that the owner of the Coca-Cola plant in Uraba, Colombia, used paramilitaries to intimidate workers to leave trade unions and to accept poorer conditions.

Coca-Cola were acquitted of allegations in US courts but worrying rumours continued to circulate on TV and on social media.

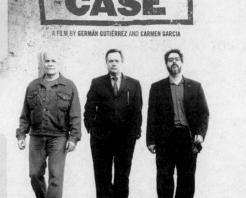

THE COCA-COLA CASE

A FILM BY GERMÁN GUTIÉRREZ AND CARMEN GARCIA

THE TRUTH THAT REFRESHES!

Figure 2 ◗
Coca-Cola Case documentary film

	Social	Economic	Environmental
Positive	Training and education programmes (e.g. 5by20 programme to enable **empowerment** of female entrepreneurs).	Franchise operation means that many local bottlers profit from sales.	Uses marketing network to increase awareness of recycling and distribution network for disaster relief.
	The Coca-Cola Foundation awards grants to companies throughout the world.	Investment in new plants in expanding markets (e.g. China and North Africa).	Initiates sustainable agricultural schemes (e.g. tea suppliers in China).
	Employment opportunities, directly and indirectly via related industries.	Investment in new markets drives economic growth.	Replenishes the water it uses (e.g. by funding local projects to protect watersheds).
Negative	Harsh working conditions in some bottling plants.	Long hours for little pay.	Exhaustion of local water supplies.
	Millions spent countering links to obesity (e.g. European Hydration Institute).	Majority of profits are returned to shareholders back in USA.	Water pollution (e.g. Kerala, India 2004).
	Workers encouraged to abandon union membership in some LDEs.	Vulnerability of bottling plants to the effects of top-down decision-making.	Pesticide residue identified in some Coca-Cola products.

▲ **Figure 1** *Impact of Coca-Cola on host countries*

Sixty second summary

* Coca-Cola is one of the world's most successful and recognisable brands; it is an example of a global product.
* There are 20 main Coca-Cola brands that are sold in nearly 200 countries.
* There are a range of social, environmental and economic impacts of Coca-Cola's business.

Over to you

A staggering 94% of the world population recognise the Coca-Cola brand. From memory, write down the characteristics of a global product.

Student Book
pages 30–3

You need to know:

- about world trade in bananas
- about the growth and benefits of Fairtrade.

What is Fairtrade?

The purpose of **Fairtrade** is to create opportunities, through sustainable development, for producers and workers who have been economically disadvantaged or marginalised by the conventional global trading system. Fairtrade aims to:

- pay farmers a guaranteed minimum price
- offer fair trade terms
- reinvest in the local community.

🔺 **Figure 1** *The first Fairtrade label was launched in the Netherlands in 1988*

Big idea

Fairtrade helps producers and workers who are disadvantaged by the global trading system.

Banana trade

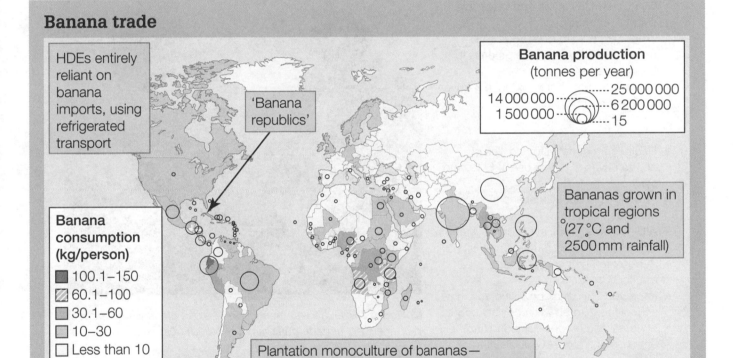

HDEs entirely reliant on banana imports, using refrigerated transport

'Banana republics'

Banana production
(tonnes per year)

14 000 000 ┄┄ 25 000 000
1 500 000 ┄┄ 6 200 000
┄┄ 15

Bananas grown in tropical regions (27 °C and 2500 mm rainfall)

Banana consumption (kg/person)

- 100.1–150
- 60.1–100
- 30.1–60
- 10–30
- Less than 10
- No data

Plantation monoculture of bananas— typically owned by large fruit-exporting TNCs

🔺 **Figure 2** *Areas of banana production and consumption*

Banana TNCs

Bananas are traded globally on a vast scale. TNCs own vast plantations and have a virtual monopoly of the market. They tend to have the following characteristics:

- Initial high capital investment due to large-scale clearance of natural vegetation and installation of infrastructure (to allow access to ports for export).
- Poor environmental record, e.g. heavy use of pesticides and unsustainable cultivation methods.
- Traditionally poor pay and conditions for local workers.

- Strong vertical and horizontal integration – the same TNC chemically treats, ships and processes.
- So-called 'Banana republics' became entirely reliant on banana trade.
- Profits along the supply chain are squeezed by TNCs and also supermarkets, who have huge buying power and negotiate prices down.
- There has been a recent shift within retail sector in the UK, towards smaller and independent producers – 1 in 3 bananas sold in the UK is now Fairtrade.

Continued over >>>

How has Fairtrade grown?

Fairtrade has allowed small-scale producers to compete with large TNCs and sell their products all over the world.

- Teamwork – similar minded organisations (including many NGOs worldwide) work together to meet shared aims. There are now four international Fairtrade networks (FLO, WFTO, NEWS, EFTA).
- Government support – emergence of development politics in 1960s with its encouragement of 'trade not aid'.

- Established record of success – handicrafts were first sold in US in the late 1940s. Foodstuffs followed in the 1970s (initially coffee) and were successfully marketed and sold via 'world shops' such as Oxfam. The brand continues to grow.
- Consumer awareness – the Fairtrade brand is now widely recognised and its values understood. The brand has gained further credibility with the creation of an international monitoring system (FLO-CERT) that monitors and certifies the producers.

What are the benefits of Fairtrade for the producers?

The El Guabo Association of Small Banana Producers in south-west Ecuador now benefits both economically and socially.

Economic

- Stabilised incomes and improvements to standards of living.
- Guaranteed fair wage and long-term supply contract.
- Producers able to raise additional capital for reinvestment, e.g. tanks to wash bananas.
- Migrant labourers are helped, e.g. given assistance to buy their own land.

Social

- Health care benefits for families of cooperatives, e.g. free use of El Guabo clinic.
- Provision of educational and medical supplies.
- Affiliation with a social security system, e.g. payment of retirement benefits.
- Support for the poorest groups, e.g. food baskets.
- Improved education provision, e.g. new school for children with special needs.
- Marginalised groups helped to find employment, e.g. HIV/AIDs sufferers.

 Figure 3 *El Guabo, formed in 1997, is one of the world's largest producers of Fairtrade bananas; around 30 000 boxes a week are exported to USA and Europe*

Sixty second summary

- A relatively small number of powerful TNCs have traditionally controlled most of the banana market (from production to sale).
- Fairtrade aims to pay farmers and guarantee a fair price.
- Fairtrade bananas are now increasingly sold in supermarkets leading to a reduction of influence of TNCs.
- Associations, such as El Guabo, look after the social and economic needs of the producers and their families.

Over to you

Do a SWOT (Strengths, Weaknesses, Opportunities, and Threats) analysis of Fairtrade with reference to world trade in bananas.

Student Book
pages 34–9

You need to know:

- about the geographical consequences of global food systems
- the global impacts of palm oil trade.

Global food systems

Global food TNCs are amongst the most powerful on the planet – they impact our lives by influencing what we eat and even how we live.

Obesity is a global problem – but so too is hunger. TNCs are responsible for 'supplying' food to the nearly 2 billion people who are overweight whilst, arguably, overlooking the needs of the 800 million who are undernourished.

However, it is illogical to blame food TNCs for making us 'fat' – it is our choice to purchase the more convenient cheap, processed foods which can not only affect our health but also maintain large profits of TNCs.

TNCs, food and technology

- Food TNCs are largely responsible for controlling the global food system and for the prices that we pay for our food.
- In general, food is manufactured cheaply in order that TNC profits are maximised. Cheap, processed food is invariably unhealthy.
- Since 2015, genetically modified (GM) crops have begun to enter the UK food chain (mainly via animal feed) although widespread planting of GM crops is low when compared to other HDEs.
- Artificial additives extend shelf life (and help to retain the 'aesthetic qualities' of food).
- As a result of invention and **agrotechnologies**, sufficient food is grown to feed all of the global population.
- In 2014, almost half of the grain harvested worldwide was not processed for human consumption – it was used as animal feed, biofuels or raw materials for industry.

 Big idea

The global food system includes the growing, harvesting, packing, processing, packing, transportation and consumption of food – a handful of TNCs have a large share of the market (Figures **1** and **2**).

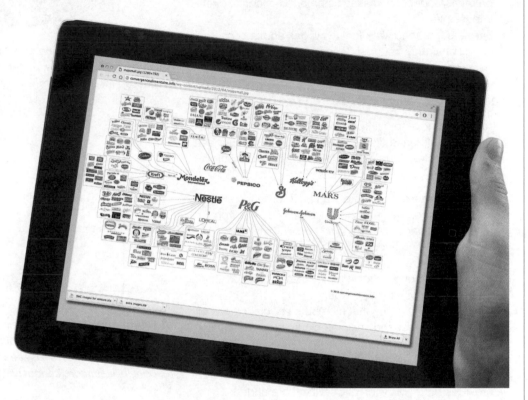

Figure 1 *Less than ten food TNCs control more than 50% of the food on sale in a typical UK supermarket*

1 Associated British Foods

2 Coca-Cola

3 Danone

4 General Mills

5 Kellogg

6 Mars

7 Mondelez International

8 Nestlé

9 Pepsi

10 Unilever

Figure 2
Just ten firms control 28% of all food production

Continued over ▶▶▶

Trade in palm oil

Palm oil is the most used vegetable oil worldwide because its high melting point makes products smooth, creamy and easy to spread. It is also cheap and widely available. As a result there is a lucrative and vigorous trade in palm oil worldwide.

Impacts of palm oil cultivation

- Rainforest is cleared, particularly in south-east Asia, to allow large-scale plantation monoculture. Clearance by burning releases greenhouse gases.
- Locals may be forced off land or have land rights challenged by palm oil TNCs.
- Water and soil pollution, caused by chemicals used in palm oil production, makes other forms of agriculture impossible.
- Farmers forced to work for low wages and in poor conditions; they become dependent on TNCs and traditions may be lost.

Sustainable palm oil

Is the answer to stop using products containing palm oil? The Roundtable on Sustainable Palm Oil (RSPO) suggests not and promotes sustainable methods of production of palm oil.

- The RSPO represents more than 2000 producers, processing companies, retailers, banks, NGOs from over 75 countries.
- To be RSPO-certified, palm oil must be produced where there is environmental conservation, consideration of needs of employees and responsible development of new plantings.
- Critics of RSPO argue that members still clear areas of pristine rainforest and, in some plantations, the working conditions remain poor.
- GreenPalm certificates are issued to producers and plantations that meet RSPO standards. This provides traceability for TNCs, which are under pressure to prove their green credentials.
- There are ongoing concerns for the tropical rainforest biome, in particular for populations of endangered orangutan.

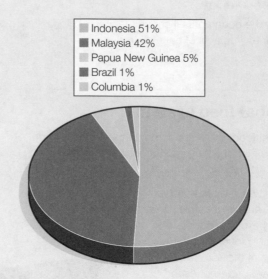

Indonesia 51%
Malaysia 42%
Papua New Guinea 5%
Brazil 1%
Columbia 1%

 Figure 3 *Consumption of palm oil (thousand tonnes), which continues to grow, matching our hunger for processed food products*

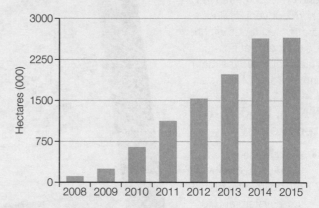

Figure 4 *Total area of sustainable palm oil production. Increasing public awareness of sustainability, have encouraged producers to take their environmental responsibilities more seriously*

Sixty second summary

- Less than ten TNCs control half the food on sale in a British supermarket.
- The amount of food being produced worldwide has increased due to agrotechnologies.
- There has been a large increase in demand for palm oil.
- The palm oil trade has positive and negative local and global impacts.
- There is a need for palm oil production to be more sustainable.
- GreenPalm certificates provide traceability for TNCs.

Over to you

It is important to be able to use the term 'sustainability' correctly. Mind map the term 'sustainability'. Try and make reference to global food systems.

Student Book
pages 40–1

You need to know:
- what global governance is
- about norms, laws and institutions.

What is global governance?

The Commission on Global Governance describes global governance as the '... *sum of the many ways individuals and institutions, public and private, manage their global affairs ...*'

A range of **actors** manage their global affairs, impacting and contributing to the shaping of wider global systems. They act both *reactively* and *proactively* and include:

- nation states
- profit-making companies – especially TNCs
- NGOs
- international institutions, such as the UN.

Global governance is, therefore, an international process of consensus-forming which, in turn, generates guidelines and agreements 'governing' the actions of those same players.

Norms and laws

Norms, rules and laws are constantly made and remade which shape global systems:

- Norms are the values, traditions and customs that govern behaviour in societies, e.g. the need to keep people safe at work. Norms often become laws.
- Not all societies agree as to what is normal and reasonable, making global governance (and the role of institutions such as the UN) very complex.
- Interpretation and enforcements of norms and laws can differ between signatories of international agreements and treaties. This can have consequences in areas such as the protection of workers' rights (Figure **1**).

See Figure 1 on page 40 of the student book for more detail on these actors.

Figure 1 *Our right to work in a safe environment is a norm in the UK that we take for granted. Even where rules about the need for safe working conditions are accepted they may not be enforced – as the collapse of the Rana Plaza building in Bangladesh in 2013 demonstrated.*

The UNI Global Union, in alliance with other leading NGOs, helped to publicize the plight of relatives of those crushed in this garment factory. The work of these organisations turned this event into an issue on the global stage and, eventually, led to the legally-binding *Bangladesh Accord on Fire and Building Safety*. Though increasing numbers of TNCs have now signed the accord, only time will tell how effective it is.

In a globalised world, transnational problems need solving but, perhaps, only international institutions, such as the UN, WHO, World Bank and the International Criminal Court, are in a position to do this. For example, flows of refugees, workers' rights, financial market instability, **trade protectionism** and climate change all need regulation, monitoring and control, on a global scale.

 Sixty second summary

- Global affairs are managed by a range of actors, including nation states, TNCs and institutions.
- These actors in global systems take a range of proactive and reactive roles.
- Norms, rules and laws are constantly made and remade, which shape and regulate global systems.
- Transnational problems that need addressing are the focus of international institutions such as the UN and the WHO (see 1.12).

Over to you

Practise defining the terms *global governance, actors, norms* and *institutions* in relation to global systems, in **30** words or fewer.

You need to know:

- about issues associated with attempts at global governance
- inequalities and injustices in global governance.

Student Book
pages 42–5

Sovereignty and global governance

Your atlas shows nearly 200 individual sovereign states – each with their national territories and security responsibilities. But, exclusive sovereignty is not realistic in an increasingly interconnected, globalised world. Just as rivers do, people, money, ideas and information leak out of individual nations and flow across borders into other countries.

In reality, nation states:

- exercise sovereign authority over their territories and interests beyond national boundaries
- 'pool sovereignty' when it is in everyone's best interests (e.g. collaboration in military intelligence, border security and joint missions in organisations such as NATO).

Issues associated with attempts at global governance

Deepening economic globalisation, as well as increasing migration, trade and capital flows, make individual states more and more susceptible to policies adopted by others (see 1.6).

Likewise, climate change and increased activities in the global commons (see 1.13) do not fall within the jurisdiction of any one particular country. This is why organisations such as the WTO, WHO and the **United Nations Environment Programme (UNEP)** are championed by many to take the lead in trade, health and environmental issues.

Big idea

If the world had a global government, it would be called the United Nations.

🔺 **Figure 1** *One world?*

The United Nations – a global government?

Arguably the most globally significant sets of norms, laws and institutions arise from the work of the United Nations. The UN was established in 1945 immediately after the Second World War to maintain world peace. It has since developed and subsequently enforced international laws and policies that impact directly or indirectly on every one of the world's citizens.

Article 1 of the UN Charter states its purposes, summarised to:

- maintain international peace and security
- develop friendly relations among nations
- promote equal rights and self-determination of peoples
- achieve international cooperation
- encourage respect for human rights and fundamental freedoms.

🔺 **Figure 2** *The UN Headquarters Secretariat building in New York City; if there was a global government, it would be the United Nations*

📖
Figure 3 on page 44 of the student book has a full summary of the work of the UN.

Inequalities and injustices in global governance

The Centre for International Governance Innovation (CIGI) has identified a number of inequalities and injustices ('gaps') in almost every sector of global governance. They all impede progress in the global economy, security, development and the environment. These gaps refer to:

- *Jurisdiction* – the gap between a need for global governance in many areas (e.g. water security) but the lack of an authority with the power, or jurisdiction, to take action.
- *Incentive* – the gap between the need for international cooperation and the motivation to undertake it. This gap is closing as globalisation provides increasing impetus for countries to cooperate.
- *Participation* – the gap acknowledging that international cooperation remains primarily the affair of governments, leaving civil society groups on the fringes of policy-making.

Examples of CIGI concerns include:

- environmental governance involving so many agencies and agreements that duplication and incoherence is restricting progress
- the need for collaboration to improve among the IMF, G-20 and OECD if international jurisdiction and regulation of finance is to improve
- the need for an international law to ensure and enforce water security
- the need for better coordination to deal with epidemics and vaccine stockpiling.

▲ Figure 3 *Chinese President Xi Jinping addressing the UN General Assembly in 2015.*

Troubling gaps were also highlighted in the global governance of science and technology, urbanisation, migration, energy and even trans-national crime (e.g. cybersecurity).

With great power comes great responsibility

China has changed the way it does business and, as its wealth and power grow, so also will its interests expand (Figure **3**).

With its increasing influence on the world stage, China may feel the need to intervene in the Middle East peace process. How will the contribution of this new superpower be received then?

 Sixty second summary

- In an increasingly interconnected, globalised world exclusive national sovereignty is no longer realistic.
- In almost every sector of global governance, a number of critical inequalities and injustices impede progress.
- These 'gaps' cover three key themes – jurisdiction, incentive and participation.
- Of all organisations central to global governance, the UN is best placed to lead, guide and coordinate productive responses to transnational problems.

Over to you

Make sure that you appreciate, and can state examples of the perceived **strengths** and also **weaknesses** of the UN as an agent of global governance.

You need to know:

- about global commons – the concept, rights and benefits
- the way the protection of the global commons is key to sustainable development for all.

Student Book
pages 46–7

Global commons – the concept

Global commons are defined as those parts of the planet that fall outside national jurisdictions and to which all nations have access. These are:

- the high seas and deep oceans
- the atmosphere
- the northern and southern polar regions – Antarctica in particular
- outer space.

Some widen the concept to include resources of interest or value to the welfare of all nations, such as biodiversity. Indeed, the global commons may well be '*the collective heritage of humanity – the shared resources of nature and society that we inherit, create and use*'.

Global commons – rights and benefits

Rights and benefits associated with the global commons can only be assured through effective management on a global scale. A strong and effective UN could ensure the coordination, cooperation and coherence required.

The new UN Sustainable Development Goals (Figure **2**), adopted in 2015, could provide a solution – a perfect vehicle for promoting and enabling:

- the formulation of rules and regulations governing use of the global commons
- the monitoring and enforcement of those rules.

Current conventions and treaties governing the global commons include:

- *High seas*: 1982 United Nations Convention on the Law of the Sea (UNCLOS III).
- *Atmosphere*: United Nations Framework Convention on Climate Change (UNFCCC).
- *Antarctica*: Antarctic Treaty System (ATS).
- *Outer space*: Treaty on Principles Governing the Activities of States in the Exploration and Use of Outer Space.

 Figure 1 *Outer space, the poles, oceans and atmosphere are global commons*

 Figure 2 *Sustainable Development Goal 14 is to* 'conserve and sustainably use the oceans, seas and marine resources for sustainable development'.

Sixty second summary

- Global commons are defined as those parts of the planet that fall outside national jurisdictions.
- They include the high seas and deep oceans, the atmosphere, the northern and southern polar regions and outer space.
- A number of conventions and treaties oversee their governance, but they are difficult to enforce and by no means comprehensive.
- The UN's Sustainable Development Goals offer a new opportunity to promote the necessary protection of the global commons.

Over to you

Why was the inclusion of the protection of the global commons within the UN Sustainable Development Goals crucial?

You need to know:
- about the contemporary geography of Antarctica
- Antarctica's fishing, whaling and mineral resources.

Student Book
pages 48–51

Antarctica – the world's last great wilderness

Antarctica is a unique continent of extremes:

- If it were a single country, only Russia would be larger in area.
- It is almost entirely covered in ice, which flows out to sea as slow-moving ice shelves and glaciers.
- It has no native (indigenous) population, but surprising biodiversity.
- It is rich in mineral resources.
- It is home to around 4000 scientists in over 50 research stations.

Beneath 14 million km² of Antarctic ice (4 km thick in places) lies a continent formed of two 'blocs' separated by a deep channel. Tectonics explains that East Antarctica shares similar ancient rocks to South Africa and Australia (exposed at the surface in the Transantarctic Mountains).

Beneath the ice of West Antarctica is an **archipelago** of steep mountainous islands, much of which is exposed in the Antarctic Peninsula.

Where Antarctica's rocks are exposed, vegetation is limited by cold, aridity, wind and lack of soil. Lichen and moss predominate – the latter particularly in the peninsula area where summers are warmer and hardy flowers can grow. In contrast, the surrounding seas support a rich variety of marine birds (e.g. penguins and petrels) and seals which feed at sea, breed on the land or sea ice, but rarely venture inland.

The seas around Antarctica:

- are frozen or covered with pack-ice for much of the year
- have a moderating influence on coastal temperatures (compared with the continental interior)
- are naturally bounded by the Antarctic Convergence.

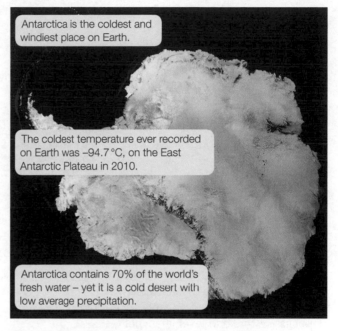

Antarctica is the coldest and windiest place on Earth.

The coldest temperature ever recorded on Earth was –94.7 °C, on the East Antarctic Plateau in 2010.

Antarctica contains 70% of the world's fresh water – yet it is a cold desert with low average precipitation.

Figure 1 *Antarctica from space*

The Antarctic Convergence

Delimiting Antarctica is pertinent to its protection. The Antarctic Treaty System covers the area south of 60°S latitude, but there is also a looping, natural boundary called the *Antarctic Convergence*. This watery dividing line, up to 48 km wide in parts, separates cold north-flowing waters from the warmer waters of the sub-Antarctic. Along it there is significant mixing and upwelling, creating a highly productive marine environment.

Figure 2 *Adelie penguins on an Antarctic iceberg*

Continued over ▶▶▶

Fishing and whaling

Following its 'discovery' in the eighteenth century, economic exploitation of Antarctica has focused on the resources of the surrounding seas.

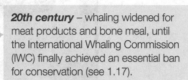

18th century – hunting of fur seals to the point of virtual eradication.

19th century – Norwegian, British and American whaling of blue and right whales for oil and baleen (whalebone).

20th century – whaling widened for meat products and bone meal, until the International Whaling Commission (IWC) finally achieved an essential ban for conservation (see 1.17).

21st century – fishing now the main contemporary economic focus. But Russian and Japanese exploitation of the Southern Ocean for rock cod and krill (central to the whole Antarctic food web) is raising serious concerns about overfishing. Conservation is essential.

 Figure 3 *Exploitation of resources of the seas around Antarctica*

Exploitation of mineral resources

Known mineral resources in Antarctica include coal, oil, manganese, titanium and even gold and silver. But sizeable accessible deposits are rare and, even then, not yet economically viable to mine.

Any mineral exploitation in Antarctica would have to overcome the seriously hostile environment characterised by:

- major problems of inaccessibility
- the extreme and hazardous climate
- deep covering of moving ice sheet and glaciers.

 The data on page 51 of the student book gives an indication of Antarctica's hostile environment.

 Figure 4 *Abandoned whaling ship at Grytviken, South Georgia Island*

Sixty second summary

- Antarctica covers a huge area and is the coldest, windiest place on Earth.
- The seas around Antarctica are frozen or covered with pack-ice for much of the year.
- Flora and fauna is limited by cold, aridity, wind and lack of soil.
- Economic exploitation of Antarctica has focused on the surrounding seas - fishing, whaling and sealing.
- Exploitation of Antarctica's rich mineral deposits would have to overcome the hostile environment.

Over to you

State and learn at least **five** memorable facts associated with the contemporary geography of Antarctica. For example: Antarctica is the windiest place on Earth.

Student Book
pages 52–3

You need to know:
- about scientific research and threats to Antarctica arising from climate change.

Scientific research

Thirty nations run scientific research stations in over 50 widely-scattered coastal and interior locations (Figure **1**). Collectively, this represents a multinational laboratory of incalculable value, conducting research into atmospheric, terrestrial and oceanic systems. For example, ice core drilling through layers of hundreds of thousands of years of snowfall, reveals an abundance of information about past climatic conditions.

Figure 1 ❯
The British Antarctic Survey's Port Lockroy research station was established in 1944

Climate change

Antarctic research has provided the clearest link between levels of greenhouse gases in the atmosphere and global surface temperatures (Figure **2**).

Analysis of the properties of ice core layers, and also air bubbles captured within them, allows scientists to reconstruct the pattern of changing air temperatures along with past concentrations of greenhouse gases. Such analysis has revealed that:

- climates follow natural cycles throughout geological time
- climates will continue to change in the future (with or without humans)
- human activities (e.g. emissions of greenhouse gases) have contributed to contemporary climate change (global warming).

But the trends are not straightforward:

- East Antarctica's ice sheet is actually thickening – warming seas cause increased evaporation, which condenses into cloud droplets, and falls as snow in the frozen interior.
- West Antarctica's smaller, *warm-based* ice sheet (see *AQA Geography A Level & AS Level Physical Geography* 4.5) is more likely to slide into the sea than the Eastern ice sheet. This would raise global sea levels by 5 m.
- Antarctic Peninsula temperatures are rising up to five times faster than the rest of the world. Consequently, ice shelves have been breaking up, no longer holding back land-based ice. If these ice sheets slide into the sea, vast quantities of water will be added.

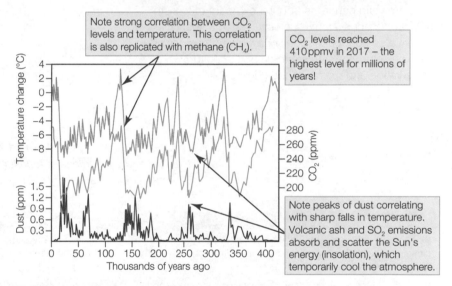

Note strong correlation between CO_2 levels and temperature. This correlation is also replicated with methane (CH_4).

CO_2 levels reached 410 ppmv in 2017 – the highest level for millions of years!

Note peaks of dust correlating with sharp falls in temperature. Volcanic ash and SO_2 emissions absorb and scatter the Sun's energy (insolation), which temporarily cool the atmosphere.

❯ **Figure 2** *The results of an ice core drilled 3.5 km down into the Antarctic ice in 1999 revealing data from four glacial cycles*

 Sixty second summary

- Scientific research on Antarctica represents a multinational laboratory of incalculable value.
- Analysis of ice cores, extracted by drilling through layers of hundreds of thousands of years of snowfall, reveals an abundance of information about past climatic conditions.
- The impact of climate change is most apparent in the polar regions, with the Antarctic Peninsula particularly sensitive.

Over to you

Outline the challenges facing scientists working in Antarctica (Twitter, blogs and the BAS website are both revealing and inspiring).

You need to know:

*Student Book
pages 54–5*

- about threats to Antarctica arising from tourism.

Increasing tourist numbers

Advances in transport, technology and clothing have allowed Antarctica to be accessible to increasing numbers of tourists – albeit limited to more affluent groups. Records of visitor numbers are dominated by groups from HDEs (Figure **1**). This is not mass tourism.

- The vast majority visit by cruise liner, seeking the chance to see vast icebergs, whales and penguin colonies.
- 'Adventure tourists' seeking photo safaris, kayaking and so on, fly to ice runways, or use smaller ships, allowing easier opportunities to land.

Impacts of tourism

There are valid concerns about the impact of tourism on Antarctic ecosystems. However, such a financially exclusive sector could equally be viewed as a powerful force for Antarctica's environmental and cultural conservation and maintenance (Figure **2**).

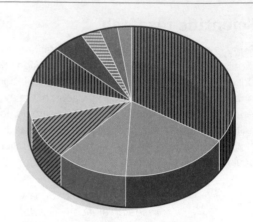

USA (12 308)

Australia (4087)

United Kingdom (3428)

China (3042)

Germany (2796)

Canada (1937)

France (1092)

Switzerland (1064)

Japan (831)

Others (6117)

Figure 1 ◗
*Number and proportions of tourists to
Antarctica by nationality (2014–15)*

Arguments for tourism	Arguments against tourism
Strict governance, self-regulation and relatively low numbers of closely managed visitors result in minimal impact – only a tiny proportion of Antarctica is affected.	Cultural heritage sites – such as old sealing, whaling and exploration stations – are under pressure and require repair.
Only 5% of landing sites show signs of wear and tear, and litter is negligible.	The peak summer (mid-November to March) season coincides with wildlife breeding periods. Given clustering of popular species (e.g. penguin colonies), disturbance may be unavoidable.
Visitors tend to be well educated, responsible, environmentally aware and expertly guided in small groups.	Antarctic ecosystems are extremely delicate and vulnerable to accidentally introduced invasive species, pollution and disturbance.
Strict regulation of marine pollution and waste discharge suggests that long-term protection is assured.	Marine safety cannot be guaranteed. The sinking of the *MS Explorer* in 2007 demonstrates just how hazardous icebergs can be.

◔ **Figure 2** *Arguments for and against tourism in the Antarctic*

 Sixty second summary

- Advances in transport, technology and clothing have led to more tourists visiting Antarctica.
- Antarctica presents incomparable opportunities for both cruise and adventure tourists to experience the trip of a lifetime, but at a price.
- Antarctic tourists tend to be from HDEs and usually wealthy, responsible, environmentally aware.
- There are risks associated with tourism to Antarctica, but not all impacts threaten the fragile Antarctic ecosystem.

Over to you

Produce a series of flashcards to summarise the impacts of tourism in Antarctica.

Student Book
pages 56–9

You need to know:
- about the governance of Antarctica
- about the Antarctic Treaty, the Protocol on Environmental Protection to the Antarctic Treaty (1991) and the IWC Whaling Moratorium
- the role of NGOs in monitoring threats and enhancing protection.

Who governs Antarctica?

Antarctica is not a nation state. But over the years, seven countries have made claims to segments of it – all of which have been disputed (Figure **1**). However, the areas assigned are recognised as research zones where individual countries run scientific bases.

'Land of science. Land of peace'

This lack of sovereignty is why the Antarctic Treaty System (ATS) is so important. In December 1959, twelve nations, including the USA and, as it was then, the USSR (Russia), signed this international agreement not to recognise, dispute, establish or allow future claims of territorial sovereignty over Antarctica. Crucially, the ATS (which now has 52 signatories) would:

- guarantee free access and research rights to all countries
- prohibit military activity, such as nuclear bomb tests
- ban the dumping of nuclear waste.

But pressure has grown to exploit Antarctica's mineral resources. NGOs, acutely aware of the potentially drastic consequences of this, and with public support globally, proposed replacing the unenforceable convention with a formal protective protocol. Hence the *1991 Protocol on Environmental Protection* which saw signatories agreeing until 2048 to:

- ban all mineral resource activity including exploration of the continental shelf
- promote comprehensive monitoring and assessment in order to minimise human impacts on Antarctica's fragile ecosystems.

NGOs in Antarctica

The work of NGOs (non-governmental organisations) is very important in:

- concentrating and providing expertise
- championing causes
- contributing independent perspectives
- rallying public support
- provoking action.

See Figure 3 on page 57 of the student book for in in-depth examination of the Protocol on Environmental Protection

> **Big idea**
>
> Concepts of sustainability and environmental stewardship have developed directly from the world conservation movement.

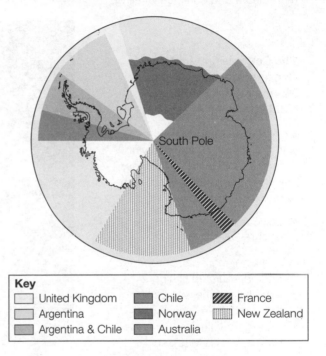

South Pole

Key

United Kingdom	Chile	France
Argentina	Norway	New Zealand
Argentina & Chile	Australia	

▲ **Figure 1** *Territorial claims to Antarctica; it is easy to see how disputes can occur*

For example, concepts of sustainability and environmental stewardship have developed directly from the world conservation movement, which includes high-profile NGOs such as the World Wide Fund for Nature (WWF).

The combined efforts of NGOs in Antarctica and the Southern Ocean – specifically the work of the Antarctic Ocean Alliance (AOA) and the Antarctic and Southern Ocean Coalition (ASOC) – has been crucial in championing the case for designating **marine protected areas (MPAs)** and **marine reserves** where wildlife is under increasing pressure from commercial fishing and climate change.

Continued over

The International Whaling Commission (IWC)

The IWC was set up in 1946 to monitor and conserve global whale stocks and oversee the whaling industry. It is credited with achieving a moratorium banning all commercial whaling from 1986. This agreement is viewed as one of the most significant success stories to date in the governance of the global commons. It remains in place today, although Norway and Iceland object to the moratorium and continue whaling.

The Commission for the Conservation of Antarctic Marine Living Resources (CCAMLR)

CCAMLR was established in 1982 with the objective of conserving Antarctic marine life – especially *krill*, a keystone component of the Antarctic ecosystem.

Specific achievements of CCAMLR include:

- challenging illegal, unreported and unregulated fishing
- establishing the world's first high seas MPA
- managing vulnerable marine ecosystems (VMEs), such as seamounts, hydrothermal vents and cold water corals by regulating bottom fishing on the high seas.

Analysis and assessment of global governance in Antarctica

For over fifty years the ATS has prevailed with no major problems – an inspiring example of international, peaceful cooperation promoting invaluable scientific research (Figure **2**).

But ATS decision-making can be problematic:

- Decision-making by consensus does not mean that everyone must agree, but that no one can voice disagreement – so resolutions can be stalled if one country feels strongly about an issue.
- Without legal penalties most parties are essentially on their honour to abide by their obligations.

Mutual trust clearly works in Antarctica and, with the support of NGOs, much is agreed and achieved. The protection of this remarkable global common is moving forward.

▲ **Figure 2** *Flags of the founding members of the Antarctic Treaty; the symbolic 'barber pole' topped with a shiny metallic sphere marks the location of the South Pole*

- Antarctica is not a nation state, and so has no government to manage or protect it.
- The inspirational Antarctic Treaty System was signed in 1959.
- The 1991 Protocol on Environmental Protection secured key protections.
- Environmental NGOs, whether grouped or individually play an important role in protecting Antarctica.

Over to you

Make sure that you understand the significance of specific clauses of the 1991 Protocol on Environmental Protection (see Figure 3 on page 57 of the student book).

You need to know:
- about the arguments for and against globalisation.

Student Book
pages 60–1

The importance of critique

To 'critique' is to evaluate theories or practice in a detailed and analytical way. It is a challenging requirement of studies at this level and requires:

- secure knowledge and understanding of the issue(s)
- the confidence to express an opinion, yet appreciate and articulate alternative points of view
- practice.

Globalisation is a controversial topic. Diverse views about its nature, trajectory and impact are held on both sides of the political spectrum. Your capacity to *critique* is likely to be tested with regard to this topic.

Figure **2** summarises a range of key arguments used for and against the process of globalisation.

▲ *Figure 1* *Dimensions of globalisation? Powerful imagery demanding a response.*

For globalisation	Against globalisation
It reduces poverty – creating greater opportunities for all to participate and to succeed.	Inequality has increased – the richest 20% of the world's population consume 86% of the world's resources.
It enables free trade, removing barriers such as tariffs and subsidies, which helps LDEs.	Many barriers to trade still exist (e.g. 161 countries have VAT on imports).
Competition between countries drives prices down.	Countries manipulate their currency to obtain a price advantage.
Interconnected geopolitical decision-making benefits all (e.g. NGOs operate globally).	HDE governments exercise disproportionate influence in key international organisations.
The world is a safer place.	Global trade in military hardware is big business.
Cultures are more widely shared, leading to greater tolerance and understanding.	Homogenisation and a global monoculture have led to a dilution of local cultures.
Environmental concerns are more effectively addressed as nations work together (e.g. the ATS).	Achieving a global consensus on tackling climate change has been frustrating.
The internet has facilitated mass communication and supported economic development in LDEs.	The internet has provided a medium for sharing extremist propaganda and hatred.
TNC investment in LDEs has provided a route out of poverty.	TNCs have exploited the frailties of weaker governance in LDEs to their own advantage.
TNCs create employment, stimulate economic growth and redistribute wealth around the world.	Jobs transferred from HDEs to lower-cost economies may lead to unemployment in the host country and exploitation in LDEs.

▲ *Figure 2* The arguments for and against globalisation

 Sixty second summary

- Critique requires knowledge, understanding, analysis and assessment of issues.
- Some argue that globalisation promotes economic growth and development for all.
- Others suggest that globalisation has failed to be inclusive, resulting in populist movements which reject the political establishment (e.g. supporting Brexit and electing Donald Trump as the US president).

 Over to you

Practise outlining and exemplifying **three** strong arguments both **for** and **against** globalisation.

2 Changing places

Your exam

(AL) *Changing places* is a **core topic**. You must answer **all** questions in Section B of Paper 2: Human geography.

Paper 2 carries 120 marks and makes up 40% of your A Level. Section B carries 36 marks.

(AS) *Changing places* is a **core topic**. You must answer **all** questions in Section A of Paper 2: Human geography and geography fieldwork investigation.

Paper 2 carries 80 marks and makes up 50% of your AS Level. Section A carries 40 marks.

Specification subject content (specification reference in brackets)

Either tick these boxes as a record of your revision or use them to identify your strengths and weaknesses

Your revision checklist

Section in Student Book and Revision Guide	1	2	3	Key terms you need to understand. Complete the key terms (not just the words in bold) as your revision progresses. 2.1 has been started for you.
The nature and importance of places (3.2.2.1)				
2.1 The highs and lows of place				*place, placelessness, clone town, tourist gaze*
2.2 Defining place (and identity)				
2.3 Categories of place				
2.4 What shapes the character of places?				
Changing places – relationships, connections, meaning and representation (3.2.2.2)				
Relationships and connections (3.2.2.2.1)				
2.5 The dynamics of change				

Meaning and representation *(3.2.2.2.2)*			
2.6 Management and manipulation of place-meanings			

Quantitative and qualitative skills *(3.2.2.3)*			
2.7 Analysing different representations			
2.8 Using geospatial data in your place studies			

Place studies *(3.2.2.4)*			
2.9 Great Missenden: connected but not protected			
2.10 Great Missenden: '… not some Constable country'			
2.11 Using oral sources in your place studies			
2.12 Detroit: boom and bust			
2.13 Detroit: a home for racial segregation?			

You need to know:
- about how humans perceive, engage with and form attachments to places
- that place-meanings are bound up with different identities, perspectives and experiences.

Student Book
pages 66–7

What do we mean by place?

Geographers are interested in places in urban and rural settings, and at a range of different scales:

- house
- street
- locality or neighbourhood (see 2.9–2.13)
- city.

The tourist gaze

The **tourist gaze** is organised by business entrepreneurs and government officials, and consumed by the public (see 2.6). Even 'death sites' like Ground Zero (Figure **1**) are marketed and managed (curated) by tourism professionals, who:

- choose what visitors are allowed to access, within it
- mediate our experience of the place – for example, via guided tours, information boards, maps etc.

However, it is also worth noting that such places have a different meaning and importance for different people. Each visitor will perceive the location differently, at least in part, according to their own upbringing and prior experiences (religious beliefs, moral code, ethnicity, sexuality, education, experience of disability). (Also see 2.2, 2.9.).

Sometimes, people's differing perceptions of a place can lead to conflict.

Big idea

More than its physical location, a **place** is space given meaning by people.

Placelessness

Some places suffer from **placelessness**. This is the idea that a particular location (e.g. an airport terminal) 'could be anywhere' because it lacks uniqueness.

Human geographers propose that this occurs when global forces have a greater influence on shaping a place than local factors. British high streets are increasingly criticised for their uniformity where chain stores predominate.

Is your place a *clone town*?

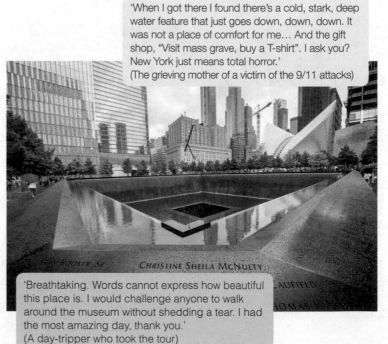

'When I got there I found there's a cold, stark, deep water feature that just goes down, down, down. It was not a place of comfort for me… And the gift shop, "Visit mass grave, buy a T-shirt". I ask you? New York just means total horror.'
(The grieving mother of a victim of the 9/11 attacks)

'Breathtaking. Words cannot express how beautiful this place is. I would challenge anyone to walk around the museum without shedding a tear. I had the most amazing day, thank you.'
(A day-tripper who took the tour)

Figure 1
The 9/11 Memorial, New York; people's experiences of the same place may be different, according to its meaning for them

Sixty second summary

- Tourist experiences of places of interest are mediated by professionals in the tourist industry; even 'death sites'.
- All places have a differing meaning and importance for individuals; they perceive them differently according to their own experiences.
- Placelessness, or increased uniformity, has been identified as a problem on Britain's high streets.

Over to you

Summarise the impact of different kinds of upbringing and experience on people's perceptions of a place you know well. Can you identify **two** groups whose perception of the same place contrasts completely?

Student Book
pages 68–71

You need to know:

- about the concept of place
- its importance in human life (the mutual links between place-meaning and identity)
- insider and outsider perspectives on place.

The concept of place

The geographical concept of place has *three different* aspects:

- **Location** – where a place is on a map, its latitude and longitude.
- **Locale** – a setting in which everyday activities (work, leisure and family life) take place. It may not be a physical place, e.g. an internet chatroom. We behave in a certain way in a specific locale, according to social rules we understand.
- **Sense of place (place-meaning)** – the subjective (personal) and emotional attachment to place, its meaning.

Figure 1 *'To the young child the parent is his primary "place".' (Yi-Fu Tuan, 1977)*

The importance of place in human life and experience

Attachment, home and identity

The depth of feeling (attachment) we have for a place is influenced by the depth of our knowledge and understanding of it – thus attachment increases with age, as:

- our physical ability to explore our environment improves
- we learn more about it.

Research also shows that our attachment to a place is influenced by the quality or intensity of experience we have there (Figure **2**).

Identity and place

Our sense of (a) place (the meaning we give to a location) can be so strong that it is central to our **identity**.

However, our attachment to place may be multi-faceted, creating a complex identity:

- county or region – 'A Yorkshireman'
- ethnicity – 'Indian'
- nationality – 'British'
- supra-national/European – 'I voted Remain.'

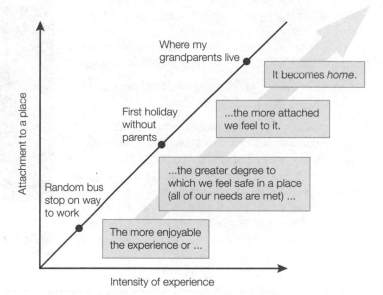

Figure 2 *Our attachment to a place is influenced by the quality or intensity of experience we have there*

Changing places, changing identities

Social, economic or political change in a location can affect the identity of the local population – such as changes as a result of Brexit (see 1.6), increased immigration (see 2.4) or HS2 (see 2.9 and 2.10). Such changes may result in a community's increased awareness of themselves as a 'people', living in a particular place.

Consciousness of, and loyalty to a place (known as localism or '**NIMBYism**'; regionalism; nationalism or patriotism; or pro-Europeanism according to the scale involved) may lead to some form of activism (see 2.5, 2.10 and 2.13).

Continued over ▶▶▶

Insider and outsider perspectives on place

All places are shaped by society: people and their regular, everyday social interactions. It follows that anybody whose behaviour varies from 'the norm' may feel uncomfortable. The dominant groups, who have the economic, social and cultural power in a location or within a society, may make them feel **out of place**.

Particular groups in society are excluded spatially, socially, politically and/or economically. Excluded groups in Britain today include the homeless, the elderly and disabled. Ethnic minorities may also feel excluded, for example, in some rural settings (see below).

	Insider	Outsider
Place of birth	Born in **X** or their parents were born there.	Not born in **X**, they are an immigrant and/or their parents and grandparents were immigrants.
Status (citizenship etc.)	Permanent resident. Holds a passport of country **X**. Can work, vote, claim benefits like free housing and healthcare.	Temporary visitor; holds a foreign passport and/or limited visa. May not be able to work, vote, claim benefits. May be travelling for business/in search of work, pleasure, safety.
Language capability	Fluent in the local language.	Not fluent. Does not understand local idioms (variations or slang).
Social interactions	Understands unspoken rules of the society of **X**.	Frequently makes faux paus or misunderstands social interactions.
State of mind	Safe, secure, happy – feels at home or 'in place' in country **X**.	Homesick, alienated, in exile – feels 'out of place'.

Figure 3 *The characteristics of **insiders** and **outsiders** in a place (country **X**)*

Are some Britons excluded from rural areas?

The percentage of the UK's *black or minority ethnic (BME)* population is around 10% and yet:

- the number of BME members of organisations like the Youth Hostel Association or the National Trust is very small
- in 2001, a UK National Parks visitor survey showed less than 1% of visitors were of a BME background.

The lack of participation of Britain's BME population in rural areas may be due to:

- *Historical imagery:* the media is dominated by historical images of a rural Britain that predates twentieth-century immigration, so BME Britons are not inspired to spend their leisure time in the countryside.
- *Southern bias:* most immigrants have moved into urban areas, which are now home to large populations of BME people. Those living outside the south-east of England feel little connection to images of a southern rural idyll.

The YHA actively encourages young people from black and ethnic minority groups to volunteer in the countryside.

Mosaic is a national project that builds links between black and ethnic minority communities and organisations such as the National Parks authorities and the Youth Hostel Association.

Figure 4 *Underrepresentation of Britain's BME population in its National Parks is being addressed*

 Sixty second summary

- The concept of place has three different aspects: location, locale and sense of place (place-meaning).
- Our understanding of, and therefore our attachment to, our environment expands with age.
- The central importance of our home informs our identity.
- Any change (or threat) to that place may lead to greater consciousness of it, and may even prompt action demonstrating loyalty; variously known as localism, regionalism or nationalism.
- Many groups in society are actively excluded in Britain today.
- Some BME Britons feel out of place in rural areas, in part, as a result of historical representations of rural Britain in the media.

Over to you

Could you give examples of localism or regionalism? If not, use 2.9 and 2.10 to find out about the reaction of a local community, living in the Chilterns, to the proposal of HS2.

Student Book
pages 72–5

You need to know:

- about near and far places
- about experienced places and media places.

Near and far places

Anthropologists, who investigate the customs and cultures of human communities, have found that everyone, wherever they live, recognises the division between 'us' and 'them'.

	Near	Far
The place	close-by, here, security	distant, 'over there', different
The people	'like us', neighbours, similar	**other**, foreign, alien, exotic

Figure 1 *The different ways we perceive near and far places and people*

National identity, difference and xenophobia

The idea that 'we are from here' and 'they are from there' is universal. It is particularly reflected in language, such as words used to describe:

- the people, e.g. *whingeing Poms* (Australian term for the British)
- different versions of familiar things, e.g. *French cricket*.

The phenomenon of perceived distance between 'us' and 'them' and between places that are **near** and **far**, has prompted a wide range of different human behaviour – from the use of mildly mocking slang terms at one end of the spectrum, to racially motivated hate crime at the other.

On the international stage, racist ideologies have been used to justify atrocities committed in wartime or under colonial powers, including the British.

Fairtrade: a different approach to the 'other'

Rather than seeing people living in distant places as 'other', with the (potentially exploitative) associated actions that follow from this *psychological construction* of distance and difference, the Fairtrade movement (see 1.9) has taken a different approach. Its aim has been to reduce inequalities between 'us' and 'them', by approaching all growers and producers with equal respect, in terms of:

- *trade* – fair prices, longer-term contracts and additional Fairtrade premium payments (to benefit farming communities)
- *representation* – case studies of farmers in distant places have been used to market products, presenting the human story behind hitherto faceless commodities, such as coffee and cocoa.

 Big idea

'Geography is about **space** ... space is the dimension that presents us with the existence of the other. Space presents us with the question, "How are we going to live together?"' (Massey, 2013)

Time–space compression

Our understanding of what is *near* and what is *far* depends on:

- how we travel
- how distance is measured (in time or kilometres).

If we use a faster method of travel, or if we use the internet to maintain contact with people in distant places, this division of the world begins to break down (Figure **2**).

With globalisation, geographers propose that space is reducing in importance and that 'the near is often an expanding domain'. (Levy, 2014)

The so-called *friction of distance* has been overcome via the use of technology, among other *factors* (see 1.1).

This is simply one of a number of *dimensions* of globalisation, which we take for granted in the twenty-first century. Do you agree?

Figure 2 *Has globalisation reduced the importance of physical distance?*

Continued over >>>

Experienced places and media places

Human geographers are increasingly interested in the way media representations of places we consume, shape and reshape our understanding of the world on a daily basis.

The representations of places that feature in the media often give contrasting images to those presented by official cartography (e.g. Ordnance Survey maps) or government statistics (e.g. census data). This is because their purpose and target audiences differ (see 2.6 and 2.7).

The role of direct experience vs. media representations

Within human geography, different thinkers debate whether places really have an intrinsic, unique nature that can only be understood by direct experience, or not (Figure **3**).

Real or imagined?

Researchers suggest that some of the places that are most important to us today exist only in books, films, games (and our imagination).

It is worth noting, however, that even **media places** are often associated with a physical location – the place that inspired the story or the location where it was filmed; for example, fans of 'Hobbiton' can take a trip to the visitor attraction near Matamata, in Waikato, New Zealand.

View 1: Direct experience allows us to perceive a location's *true nature* – to get an *authentic* sense of place	View 2: Place-meanings are always personal and socially constructed (informed by media images) – there is no *authentic* sense of place
A sensory experience: experiencing a place stimulates all of the senses – as a result we develop a deeper understanding of a location.	*Politics and power*: the most widely-held place-meanings benefit, and are reproduced by, the most powerful groups in society.
Genius loci: the term means 'spirit of a place' and is used by professional planners to convey the idea that all places have a unique set of characteristics (shaped by a set of endogenous factors, see 2.4) which new developments should take account of.	*Interdependence*: media images of places (which we haven't visited) are increasingly important at a time of greater *interdependence*, brought about by the flow of products, services, people, capital and ideas around the world (see 1.1). Places are increasingly interconnected, and shaped by exogenous factors.
Persuasive writing: 'My inclination is to go if I can…' commented David Nicholls, novelist, whose book *Us* (2015) includes a fictional account of travelling around Europe; a journey which he researched large parts of in person.	*The 'information age'*: 'We live in an age in which photography rains down on us like sewage from above…' said artist Grayson Perry (2013) speaking about the multitude of media images we are presented with every day.

 Figure 3 *Two contrasting perspectives on how place-meaning is informed by direct experience or the media*

Sixty second summary

- We perceive near and far places (and people) differently.
- Such differing perceptions have informed a wide spectrum of human behaviour from the use of mocking names for different nationalities, to racist atrocities committed by colonial powers.
- Time–space compression is changing our experience of the so-called 'friction of distance'.
- Media places, which we have only seen images of or heard about, increasingly inform our everyday lives.
- Geographers have contrasting views on whether our understanding of experienced and media places is similar or different.

 Over to you

Some higher tariff questions at A Level will make links between topics. There are some obvious links between the *Changing places* and *Global systems and global governance* topics. In around **80** words, argue **for** and **against** the idea that, in the age of globalisation, it no longer matters whether a place is near or far.

You need to know:

• about endogenous and exogenous factors.

Student Book
pages 76–9

Factors that contribute to the character of places

Factors that help to shape the character of a place may originate:

• from within a locality (**endogenous factors**, Figure **1**), such as local stone which may be commonly used in the built environment (Figure **2**).
• from beyond the locality (**exogenous factors**), relating to a relationship or connection between it and other places, people or external forces, e.g. national government policy on house-building may dictate higher housing density for all new developments, over-riding the opinions of local interest groups.

 Big idea

Places are unique, a product of both endogenous and exogenous factors (relationships and connections with other places).

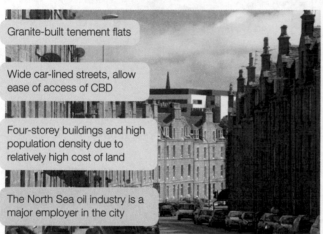

Granite-built tenement flats

Wide car-lined streets, allow ease of access of CBD

Four-storey buildings and high population density due to relatively high cost of land

The North Sea oil industry is a major employer in the city

Corallian limestone-built homes with, thatched roofs

Quaint, traditional cottages are guided by planning restrictions and tourist expectations

Narrow lanes restrict speed of traffic and resident access

A 'must see', according to international travel sites

⌃ **Figure 1** *An inner city ward of Aberdeen and the village of Abbotsbury contrast in character. Exogenous factors are in the bottom, darker boxes.*

Endogenous factors

Accents and local **dialects** vary within the UK. They:

• contribute to our understanding of the lived experience in a place
• convey a specific sense of place (place-meaning) even when we aren't there.

However, a focus on elements of local culture may also promote the stereotyping of 'the locals', hiding the diversity within the population.

Census data may arguably tell you more about a local community than analysing the dialect. Of course, social scientists draw on both **qualitative** and **quantitative** sources (see 2.11).

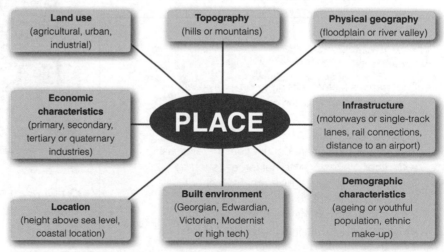

Land use (agricultural, urban, industrial)

Topography (hills or mountains)

Physical geography (floodplain or river valley)

Economic characteristics (primary, secondary, tertiary or quaternary industries)

PLACE

Infrastructure (motorways or single-track lanes, rail connections, distance to an airport)

Location (height above sea level, coastal location)

Built environment (Georgian, Edwardian, Victorian, Modernist or high tech)

Demographic characteristics (ageing or youthful population, ethnic make-up)

⌃ **Figure 2** *Physical and human endogenous factors contribute to the character of a place*

Continued over ❯❯❯

Exogenous factors

Not all influences upon places are local in origin. For example, a village may supply workers to a nearby town, or a town may be the source of day-trippers for a tourist destination.

Such influences or relationships are shown by shifting movement or flows of people (Figure **3**), resources, ideas, money and investment across space.

Migration within the EU

'Freedom of movement' is a key principle of the European Union and as a member (until Brexit, is achieved):

- the UK welcomes immigrants from the other 27 member countries
- British people are able to live and work anywhere within the EU.

Following the enlargement of the EU in 2004, flows of people into the UK from the so-called new 'EU 8' countries peaked between 2004 and 2009.

Two-thirds of 'EU 8' immigrants were Polish. Industries such as fish processing in Scotland and farming in East Anglia benefited from this influx of labour. But EU immigrants did not take up residence in an even pattern across the British Isles, and the character of some places was impacted more than others. New shops appeared on the high street (Figure **4**) and, in some areas, schools struggled to cope with large numbers of children for whom English is a second language.

Numbers of immigrants from Bulgaria and Romania increased steadily up to mid-2016, following the lifting of employment restrictions for these nationals in 2014.

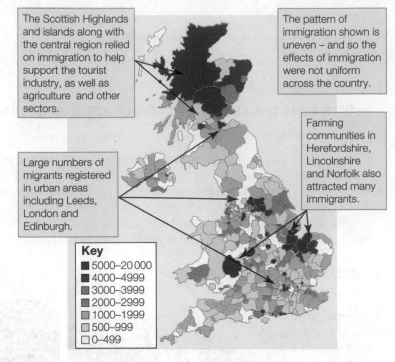

The Scottish Highlands and islands along with the central region relied on immigration to help support the tourist industry, as well as agriculture and other sectors.

The pattern of immigration shown is uneven – and so the effects of immigration were not uniform across the country.

Large numbers of migrants registered in urban areas including Leeds, London and Edinburgh.

Farming communities in Herefordshire, Lincolnshire and Norfolk also attracted many immigrants.

Key
- 5000–20 000
- 4000–4999
- 3000–3999
- 2000–2999
- 1000–1999
- 500–999
- 0–499

🔺 *Figure 3 Total numbers of Eastern European migrants registered for work in each UK local authority (2004–7)*

Polish supermarkets have become a common sight on British high streets, particularly since EU enlargement in 2004…

…but, following Brexit, for how long?

🔺 *Figure 4 The character of some British high streets reflects changing external forces*

Sixty second summary

- Endogenous factors relating to local physical (e.g. geology) and human geography (e.g. local industry, culture and dialect) help to shape the character of a place.
- Exogenous factors are connections or relationships with other places or external forces (such as regional and national government) that originate at a range of different scales.
- Exogenous factors may regulate trade, the movement of people into or out of a location and even the evolving built environment.
- The demographic characteristics of a place (using census data) helps to inform us about the character of a place.
- EU immigration into the UK, has influenced the character of places across the country, though immigrants have taken up residence in an uneven pattern.

 Over to you

Using Figure **1**, suggest how the character of the inner city of Aberdeen contrasts with the village of Abbotsbury in Dorset. Refer to factors that influence them, including what you can infer about their respective populations.

Student Book
pages 80–3

You need to know:

- about how places are socially constructed
- how past and present connections shape the characteristics of places.

Place as a social construction

In some of Asia's rapidly evolving cities, such as Shanghai in China, it is easy to see just how much has changed in recent years, especially in terms of their built environment (Figure **1**). Shanghai's connections with the rest of the world have, in recent decades, promoted significant investment in the city, a booming economy and population growth.

However, such transformations are not limited to places where skyscrapers and cranes dominate the skyline. As, in recent years, some human geographers have revisited the concept, arguing that any (and every) place is:

- *dynamic* not static
- **socially constructed**, the product of unequal power relations between people.

 Big idea

All places are *dynamic* not static, and socially constructed; often shaped by external forces.

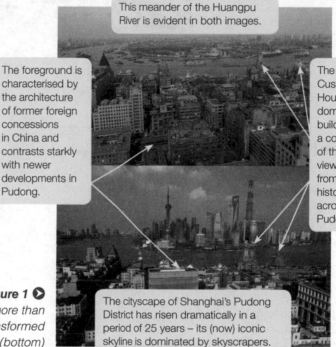

This meander of the Huangpu River is evident in both images.

The foreground is characterised by the architecture of former foreign concessions in China and contrasts starkly with newer developments in Pudong.

The distinctive Customs House and domed HSBC building allows a comparison of these viewpoints, from the city's historic centre across to Pudong.

The cityscape of Shanghai's Pudong District has risen dramatically in a period of 25 years – its (now) iconic skyline is dominated by skyscrapers.

Figure 1 ◗
Shanghai has been a hub for trade for more than 150 years; but Pudong District was transformed between 1987 (top) and 2013 (bottom)

Milton Abbey, Dorset

The UK's historic country estates may, at first glance, appear to be exceptions to the two rules stated above. However, despite appearances:

- their 'naturalistic' planting was, in fact, carefully planned (Figure **2**)
- their commanding position in the landscape may, as in the case at Milton Abbey in Dorset, be the product of the wholesale, forced movement of people off the land; who 'spoilt the view' at a particular time in history.

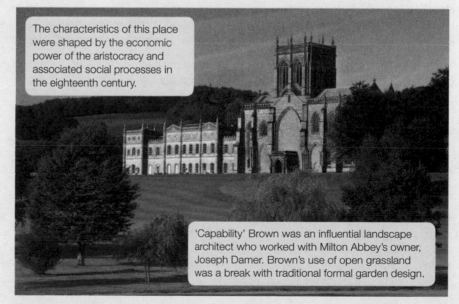

The characteristics of this place were shaped by the economic power of the aristocracy and associated social processes in the eighteenth century.

'Capability' Brown was an influential landscape architect who worked with Milton Abbey's owner, Joseph Damer. Brown's use of open grassland was a break with traditional formal garden design.

◔ **Figure 2** *Milton Abbey, Dorset: a 'powerful landscape'. Today this heritage site is a wedding venue and public school for fee-paying pupils.*

Continued over ❯❯❯

Poundbury, Dorset

Poundbury, an urban extension west of Dorchester in Dorset, is a product of a more recent *placemaking* process.

As with Milton Abbey, it is also the brain-child of an aristocrat – this time HRH the Prince of Wales.

When complete, it will contain 2500 dwellings. Work began in 1993.

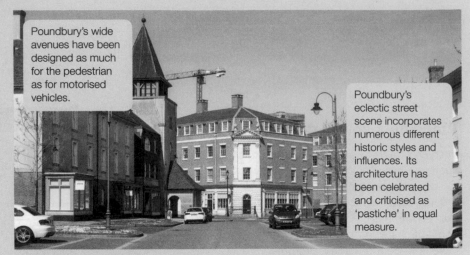

> Poundbury's wide avenues have been designed as much for the pedestrian as for motorised vehicles.

> Poundbury's eclectic street scene incorporates numerous different historic styles and influences. Its architecture has been celebrated and criticised as 'pastiche' in equal measure.

Figure 3 *Poundbury, built on Duchy of Cornwall land. Due to its innovative design, at first glance Poundbury may appear to be much older than its twenty-five years.*

Forces of change

The examples of Milton Abbey and Poundbury demonstrate that some individuals have, historically, wielded considerable power in the placemaking process – to some extent, members of Britain's royal family still do.

However, people need not be aristocracy to shape the character of places, as demonstrated by the success of community groups, such as the New Era Tenants' Association in Hackney, East London. This was a group of neighbours who worked together to resist changes in their East London estate that had been proposed by its new TNC owner, Westbrook Partners.

There are various players involved in the social process of placemaking, including corporate bodies: local government, TNCs and national institutions (Figure **4**).

Local community groups e.g. New Era Estate residents

National government

International institutions e.g. European Union

Transnational corporations (TNCs) e.g. Tata Steel, Tesco

CHANGING PLACES

Global institutions e.g. World Trade Organization

Individuals e.g activists, aristocrats, celebrities

Local government

National Institutions e.g. National Trust

Figure 4 *Forces of change that shape the characteristics of places operate at different scales, from the local to the global.*

Sixty second summary

- All places are constantly shaped (and reshaped) by social processes.
- Some social processes originate from within a locality, others original from outside.
- Social processes operate at a range of scales; from regional or national government to that of international institutions (e.g. the EU) or global organisations (e.g. the WTO).
- Past as well as present connections shape places.
- The impact of past processes may be evident in landscape design or the built environment.
- While a range of external forces may shape a place, community or local groups also have an influence; by resisting proposed changes to protect their way of life.

Over to you

Make time to revise *your* local and distant place studies. What external forces of change have shaped these places? What evidence is there of such past/present connections in the economy, the built environment or the demographic/socio-economic make-up of these communities?

You need to know:
- about how external agencies attempt to influence or create specific place-meanings
- that these external agencies include government, corporate bodies and community or local groups.

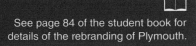

Student Book
pages 84–9

Rebranding and place-meaning

People working in the creative industries are paid by companies, councils and even national governments to help give new meaning to locations. This is **rebranding** and it is undertaken in order to boost footfall (shoppers, visitor numbers) or the employment prospects of an area, by changing our perception and subsequently, they hope, our behaviour.

Rebranding takes place at different scales and may address different audiences.

See page 84 of the student book for details of the rebranding of Plymouth.

Local scale: Llandudno

The marketing of Llandudno in North Wales as 'Alice Town' is a contemporary example of a local rebranding effort.

Inspiration and an existing place-meaning

Llandudno, 'Queen of the Welsh Resorts' was developed by the Mostyn family in the second half of the nineteenth century as a purpose-built holiday resort. The Mostyns still own much of the land today and, with the town council, strictly regulate new developments. As a result, visitors feel it has retained the air of Victoriana – but some areas are in need of regeneration.

The family of Alice Liddell, Lewis Carroll's inspiration for *Alice's Adventures in Wonderland* had a holiday home in Llandudn. Until 2009, the town was previously home to the Rabbit Hole Museum.

'Our aim was, in part, to remind local people of the history and some areas of the town they might never have visited, to help them rediscover its beauty… because it really is a wonderland' said Barry Mortlock, Director of the Alice Liddell Community Enterprise (ALICE) Ltd.

▲ **Figure 1** *The 'Alice Town Trail' and associated augmented reality apps were the idea of local entrepreneurs but backed by Llandudno town council, Conwy and Visit Wales*

A new approach

In 2012, Conwy County Council commissioned large sculptures of characters for Carroll's book, including the Mad Hatter on the promenade, to promote a literary link to this place that might otherwise have been lost. The locally-led Alice Town rebranding initiative, which capitalised on the Council's initial investment, included:

- walking trails and maps
- apps, incorporating augmented reality and digital animation (Figure **1**)
- an 'Alice Day' that begins an established Victorian extravaganza weekend in May.

The Council's larger-than-life structures were greeted with a mixed reception locally, being heralded as 'a great asset to the town' by some, while others suggested they were 'garish', 'pointless' and not in keeping with the local character.

There were, however, local benefits:

- Profits from maps and apps sold are reinvested in community projects.
- New businesses like The Looking Glass Ice Cream Parlour have created jobs in the town.

Continued over ▶▶▶

National scale: Great Mandarin names for Great Britain

See page 88 of the student book to see how communities challenge place-meanings.

Visit Britain's 'Great names for Great Britain' campaign addressed the Chinese market, in the hope of boosting overall visitor numbers but, in particular, to increase the numbers of those visiting regional attractions.

The campaign was an attempt to bridge the cultural divide, promoting Britain's regional attractions to Chinese visitors by making them more user-friendly (Figure **2**).

Client
Department for Culture, Media and Sport

Campaign budget
£1.6 million

Message
Britain welcomes Chinese tourists ('You're invited') – and now it's even easier to plan a trip as many regional attractions have Mandarin names.

Impact
- Generated new Mandarin names for 101 British attractions and boosted interest in travel to the UK.
- Estimated reach of campaign: nearly 300 million potential tourists in China in 2015.

Media
Weibo and WeChat social media platforms carried a video encouraging audience participation – making the campaign more memorable.

Opportunity:
- Chinese visitors to Britain spend £500 million a year.
- Visit Britain aims to 'double the value of that market and ensure this growth is spread across the nations and regions'.
- Chinese visitors already stay longer in Britain than in other European destinations and are high spenders.
- For every 22 additional Chinese visitors to Britain one additional UK job in tourism is created.

 Figure 2 *As part of Visit Britain's 2015 campaign, Hadrian's Wall became Yong Heng Zhi Ji (Wall of Eternity)*

 Sixty second summary

- External agencies, as well as community groups, attempt to influence or create specific place-meanings to shape the behaviour of people.
- The rebranding of Plymouth (see p84 student book) was intended to promote the city's cultural amenities, in order to retain skilled workers.
- In Llandudno local entrepreneurs helped to publicise the historic literary link with the Alice in Wonderland story. Its place-meaning was reshaped using augmented reality to increase tourism.
- Visit Britain's 2015 social media campaign aimed to make Britain's regional sites more user-friendly to Chinese tourists by giving them Mandarin names.

Over to you

'Evaluate attempts by external agencies to influence or create specific place-meanings in a location you have studied.'

Evaluate in this context means to consider or weigh up the successes and failures of such plans. Advertising or social media campaigns could be relevant or perhaps an area of your city has been regenerated or even rebuilt. *Evidence* of whether the behaviours of individuals, groups, businesses or institutions were changed is important in your answer.

Student Book
pages 90–1

You need to know:

- that places may be represented in a variety of different forms, in diverse media
- how to analyse such media images.

Critically evaluating representations of places

We rely on many different images to inform and construct the meaning we attach to places (see 2.3). We need to be able to critically evaluate a broad range of media representations (or *texts*) of a place, in order to understand the media's impact on its wider **place identity**.

Reliability

Every media source we consider is the product of an act of creative interpretation – a *secondary*, rather than *primary*, source. So how **reliable** are they? (Figure **1**).

Figure **2** includes a range of suggested activities to help you interrogate a media text or image – a representation of a place.

Positionality

As researchers, our view is subjective and our analysis of media representations will always be limited by our *positionality* – that is our own view point according to our gender, race, age, ethnicity etc.

Having an awareness of this problem, can help us to take a more reflective approach.

Figure 1 ▶

A photograph can be digitally altered to change our perception of a place

Does the source give a positive/negative impression of a place? Look for symbols or stereotypes and metaphors in the text. Think about the author/artist's choice of: • vocabulary • colour palette • camera angle.	**What is the source's provenance?** Find out about the context in which the source was produced and about its creator: • When and by whom was it produced? • What was its purpose? • Does it support or contest the views of dominant groups or powerful ideologies of the time?
How does the source compare to other available sources about the place? Further, was it produced: • earlier or later than them • inspired by them • as a reaction to them.	**How does the source relate to wider relevant geographies or processes in society?** • industrialisation • deindustrialisation • globalisation • emancipation of women.
What is the source that is being studied silent about? Look for *subtexts* or hidden texts. What was the author/artist aware of but has chosen to leave out of his or her work, such as: • men • women • economy • environment.	**How is your interpretation of the source coloured by your own experience?** Reflect on your views of the source and how it may be shaped by your: • upbringing • age • gender • the time in which you live (etc.).

▲ *Figure 2* How to interrogate a media text or image

Sixty second summary

- By analysing the content, purpose and wider context of different media representations, we may be better able to understand their message and their impact on our perception of places.
- Any attempt to analyse such 'texts' or images is limited by our own subjective view of the world – our positionality.
- However, having an awareness of the way in which your own view is limited by your experience, age, gender etc. is helpful.

Over to you

Exam questions often include an image or text you have never seen before. Write a list of **five** questions you would use to interrogate any place representation you are presented with.

You need to know:

- the value of big data, including geospatial data, and geographical information systems (GIS)
- examples of quantitative sources of data that may inform a places study.

Student Book
pages 92–5

A golden age of data?

Data about people is often geolocated (**geospatial data**), not least because of the way we interact with the World Wide Web via our phones, which incorporate Global Positioning System (GPS) technology. Such plentiful data gives an insight into the way we live and how geographic communities differ.

Big data

There are different definitions of big data but they all have several things in common:

- *High volume*: the data is not a sample, it is a record of whole datasets/population of users and it requires huge amounts of computational power.
- *High velocity*: often real-time information, e.g. purchase transactions.
- *Linked to our digital footprint*: data may be the by-product of digital interactions, e.g. the time (and our location) when we used a search engine. Concerns are being raised that our every move or browse online can be monitored. Note: data you give away has a commercial value.

Because a lot of big data has a spatial element (everything happens somewhere and often activities are geo-tagged), analysts claim to be able to use it to make predictions that relate to the population of a given area. Their predictions may allow a more cost-effective allocation of resources or, in the case of election campaigns, help a candidate to win by targeting key voters.

The UK Census

The census is the most complete source of information about the population we have in the UK. It provides a detailed picture of the entire population and is unique because it:

- surveys everyone at the same time
- asks the same questions throughout the country.

This makes it easy to compare different parts of the country.

Analysing the census

To make sense of data collected in a census, the **Office for National Statistics (ONS)** organises the responses geographically. They summarise and aggregate the data to:

- give anonymity to those who completed it – important from an ethical point of view and therefore ensure a high participation rate.
- allow wider conclusions to be drawn about populations and places at different spatial scales (Figure **1**).

Type of small area – census 2011	Average population	Avge no. of households	Total no. in E&W
Output area (OAs)	309	130	181 408
Neighbourhood or lower layer super output area (LSOAs)	1614	672	34 753
Ward (electoral districts)	6543	2726	8570

🔺 **Figure 1** *Different types of small areas are used by the ONS for detailed analysis. Data at the neighbourhood- or ward-scale is useful for a place study of a locality.*

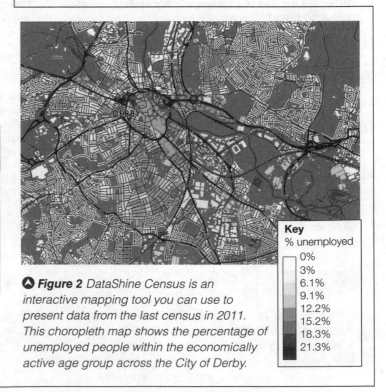

Key
% unemployed
- 0%
- 3%
- 6.1%
- 9.1%
- 12.2%
- 15.2%
- 18.3%
- 21.3%

🔺 **Figure 2** *DataShine Census is an interactive mapping tool you can use to present data from the last census in 2011. This choropleth map shows the percentage of unemployed people within the economically active age group across the City of Derby.*

The English Index of Multiple Deprivation (IMD) 2015

The English Index of Multiple Deprivation (IMD) is published by the UK Government on a regular basis and informs decision-making and associated patterns of investment. It ranks over 32000 neighbourhoods across the whole country, on the basis of seven different criteria or *domains* (Figure **3**).

The complex pattern of deprivation across England

- There are 98 neighbourhoods that have been ranked among the most deprived 1% in each IMD update (in 2004, 2007, 2010 and 2015). These are places where the community has experienced the poorest quality of life in the UK for over a decade.
- The areas most affected by persistent relative poverty are in Merseyside and Greater Manchester.
- In contrast, there were no London neighbourhoods that ranked among the most deprived 1% in each and every update.
- Some of London's boroughs were amongst the most improved between 2010 and 2015, having benefited from Olympic investment (Figure **4**).

See page 94 of the student book for a map showing multiple deprivation across England

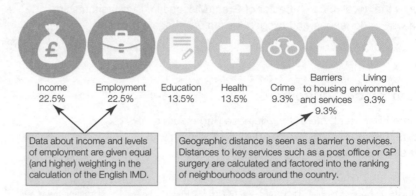

Income 22.5% Employment 22.5% Education 13.5% Health 13.5% Crime 9.3% Barriers to housing and services 9.3% Living environment 9.3%

Data about income and levels of employment are given equal (and higher) weighting in the calculation of the English IMD.

Geographic distance is seen as a barrier to services. Distances to key services such as a post office or GP surgery are calculated and factored into the ranking of neighbourhoods around the country.

> **Figure 3** The English IMD is a system of ranking neighbourhoods based on seven domains of deprivation that are combined in a weighted formula

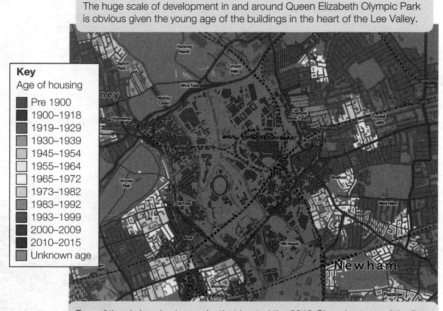

The huge scale of development in and around Queen Elizabeth Olympic Park is obvious given the young age of the buildings in the heart of the Lee Valley.

Key
Age of housing
- Pre 1900
- 1900–1918
- 1919–1929
- 1930–1939
- 1945–1954
- 1955–1964
- 1965–1972
- 1973–1982
- 1983–1992
- 1993–1999
- 2000–2009
- 2010–2015
- Unknown age

Four of the six London boroughs that hosted the 2012 Olympics topped the list of most improved in the IMD, in terms of reducing levels of deprivation 2010–15.

> **Figure 4** Age of housing mapped across the Queen Elizabeth Olympic Park and surrounding areas in East London. This Consumer Data Research Centre (CDRC) mapping tool allows the user to choose from a wide range of data sets according to their area of interest.

Sixty second summary

- Most data is geolocated, enabling cartographers and geographers to make sense of the world.
- Big data is high volume and velocity, and relates to our digital footprint – it is data we give away for free, but has a value to companies and other organisations.
- The Census provides data at a range of scales useful for a place study. It can be interrogated using online sites such as DataShine Census.
- The English Index of Multiple Deprivation (IMD) ranks more than 32000 neighbourhood areas across the country according to levels of relative deprivation (poor quality of life).

Over to you

Think about the geographical information systems that you have used to investigate different places:

- What were the advantages of investigating new locations using such technology?
- What have been the limitations of this technology for your research?

You need to know:

- about the developing character of Great Missenden, Buckinghamshire
- the way endogenous and exogenous factors shape a place.

Student Book
pages 96–9

Place Study

Dear reader, I live here

Remember that I, the author of these pages, am a white, English, educated, middle-class woman who lives in the Chilterns.

This is my *positionality* – I have a particular view on Great Missenden, shaped by my personal experience of it. You too will need to reflect on, and challenge, your own subjective perspective when investigating (and revising) the characteristics of your local **place study**.

Endogenous factors

The physical geography of Great Missenden

The village is located:

- in the Chiltern Hills, a chalk outcrop on the north-western side of the London basin
- at the head of the Misbourne Valley – the upper part of the River Misbourne flows intermittently after winter rains and dries up during the summer
- amongst ancient woodland and a network of hedgerows, which provide rich habitats for flora and fauna.

Built environment of the parish

Figure **2** illustrates the style of the built environment in the village, and the Chilterns more widely. The tradition of brick and flint construction:

- is seen as part of the 'special and distinctive character' of the Chilterns Area of Outstanding Natural Beauty (Chilterns Conservation Board, 2010)
- makes use of flints that occur in the chalk strata and the overlying clay
- often incorporates roof tiles and bricks made from the local iron-rich clay.

Changing demographic character of Great Missenden Parish

- In 1981 young dependents outnumbered elderly dependents two to one.
- Thirty years later the elderly and retired formed a larger group than those aged 15 years and under (see Figure **3**).
- According to the 2011 Census the community is not ethnically diverse: 96.3% of the population of Great Missenden is white; across England, 79.8% of the total population is white
- Great Missenden ward is among the 10% least-deprived neighbourhoods in the country
- interview data shows that a significant percentage of young people, on reaching employment age, can no longer afford to stay in the village.

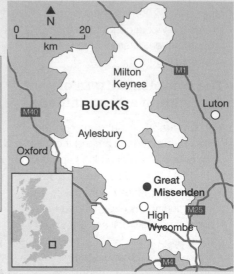

Historically, the village's location (60 km from Central London) meant it was a coaching stop, with 12 public houses and inns on the High Street. This role is still evident in the built environment.

🔼 *Figure 1* Location of Great Missenden, Buckinghamshire

🔼 *Figure 2* The built environment of Great Missenden

	1981 (%)	2011 (%)
Young dependents (0–15 years)	24.6	19.4
Working age population (16–64 years)	62.6	59.2
Elderly dependents (65+ years)	12.8	21.4

🔼 *Figure 3* Breakdown of Great Missenden population by age (Census 2011)

See page 97 of the student book for details of the Chilterns AONB and its flora and fauna.

Exogenous factors

Connected by road and rail

Great Missenden is connected to central London by road (Figure 1) and rail. It serves as a home for thousands of daily commuters who work in the capital. Prior to arrival of the railway in 1892, the village was first and foremost a farming community, with a secondary function as a coaching stop.

A place for tourism and recreation

Today, Great Missenden is at the heart of the Chilterns AONB and a honeypot site for tourists, who are drawn to:

- the natural environment of chalk hills and clay valleys, open farmland and beech woodland
- the Roald Dahl Museum and Story Centre (almost 90 000 visitors in 2016/17)
- gift shops, pubs and cafes housed in this 'quaint' village.

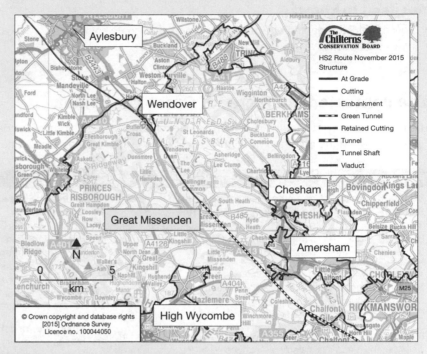

▲ **Figure 4** *The route of HS2 runs through the Chilterns AONB, and the Parish of Great Missenden*

Place Study

Faster connections and the impact of HS2

Great Missenden is undergoing economic change as a result of national government decisions to address the 'north–south divide', and influenced by the growth in construction of high-speed rail across mainland Europe since the 1980s.

- **2009:** proposal of HS2 was announced. Ministers advised that this would address capacity issues on the West Coast Main Line and reduce travel times from London to Birmingham, Manchester, Leeds and Edinburgh.
- **2010:** preferred route announced, a section of which would run through the Chilterns AONB and within a mile of the centre of Great Missenden.
- **2017:** the plan to build the first phase of HS2 (London to Birmingham) gained Royal Assent, becoming an Act of Parliament.
- **2018:** 'enabling works' began in Great Missenden (see 2.10).
- **2026:** Phase 1 due to open.
- **Up to 2033:** Phases 2a to Crewe, and 2b comprising an eastern branch to the East Midlands, South Yorkshire and Leeds, and the extension of the western branch from Crewe to Manchester. Total cost of 338 km network is estimated at £55.7bn at 2015 prices.

Sixty second summary

- The character of Great Missenden is shaped by endogenous factors, such as its physical environment of chalk-fed streams and ancient woodland.
- The built environment incorporates brick and flint construction, a local characteristic cited in the official designation of the Chilterns Area of Outstanding Natural Beauty (AONB).
- Its demographic character has changed over the last thirty years; with elderly dependents outnumbering young dependents by 2011.
- It is a place experiencing economic change due to the construction of a high-speed rail link less than a mile from the village.

Over to you

Great Missenden is a good example of a place shaped by exogenous as well as endogenous factors. Could you explain how your local place has been shaped by regional, national or international forces?

Place Study

You need to know:

- how the place-meaning of Great Missenden has become contested
- how HS2 has had an impact on the locality, to 2018.

Student Book
pages 100–1

The contested character of Great Missenden

The place-meaning of Great Missenden has become contested as part of the debate about the HS2 project. Figure **1** looks at the pros and cons of two points of view according to the framework of questions in 2.7.

Pro HS2	Anti HS2
Source B: Comment by Philip Hammond, the then Secretary of State for Transport (*The Telegraph*, 11 Dec 2010) 'Have you looked at the route? ...Between Great Missenden and the HS2 route are the A413, the Chiltern Railway and a line of pylons. So this is not some Constable country.'	**Source D: Song entitled 'Oak Tree Lament (Stop HS2)' by Dirty Mavis (2011)** 2 hundred years I stand... on the edge of the Misbourne Plain Survived the harshest winters and summers without rain My roots are deep... and you should weep For this morning strangers came, With axe and saw, and cut me to the floor So hang your head in shame
Analysis: **Positive or negative impression of this place?** Questions the alternative arguably 'bucolic' images of the AONB promoted by the Chilterns Conservation Board and local community groups at the time.	*Analysis:* **Positive or negative impression of this place?** Positive image of a natural environment, characterised by ancient woodland. It focuses on the longevity of this landscape. However, this source also presents this as a place threatened by change.
Provenance? An outsider's view: Philip Hammond did not know the area well when he made this statement – suggested by the inaccuracy about the location of the Chiltern line. As Secretary of State for Transport, his main role was to present the case for this project, hence this interview in a newspaper sympathetic to the government.	**Provenance?** An insider's view: written (after the announcement of the route) by a local musician concerned about the damage the project could do to an area he felt an attachment to; it was then adopted by the local branch of Stop HS2, a grassroots campaign.
How does it compare to other available sources about the place? Presents the 'rational' case for this project, contrasting with more emotive images in the media at the time.	**How does it compare to other available sources about the place?** It is more dramatic, arguably emotive in its presentation of the threat the project poses to the environment.
Silent about? The fact that the landscape has been designated as an Area of Outstanding Natural Beauty (AONB), a landscape of national importance.	**Silent about?** Unemployment and poverty experienced in deprived areas of the north of England, whose populations welcome this rail initiative designed to bridge the country's north/south divide.
How is your interpretation of the source coloured by your own experience? (The author writes) this statement gives an insight into the disregard the government of the time had for the views of objectors, compared to what, in their view, was its (national) importance. However, my attachment to the area makes it very hard for me to objectively appraise the minister's representation of it.	**How is your interpretation of the source coloured by your own experience?** (For the author) the personal experience of walking in the area reinforces the song's message that ancient woodland in the locality should be protected.

⬣ **Figure 1** *Analysis of Source **B** and Source **D** from page 100 of the student book*

Sixty second summary

- Contrasting representations of Great Missenden were created by different groups when HS2 was announced.
- Ministers played down the concerns of local groups about the impact of HS2, arguing that the place was of mediocre environmental quality.
- Local protestors created an image of a special and natural environment under threat.
- Great Missenden has been affected by the HS2 project including the closure of local business/services and increased anxiety amongst the local community.

Over to you

Use the questions in Figure **1** to analyse **two** qualitative sources that present contrasting images of your local place.

Jos Smith, a writer and researcher who walked the line of HS2 commented: '...these yarn-bombed trees are gestures of neighbourly concern, registering outrage but also communicating something of a defiantly humane care for a loved place. They stand to juxtapose, in this case, the lack of courtesy involved in the planning process.' (Jos Smith, 2013)

⬣ **Figure 2** *Guerrilla knitting, a protest in Great Missenden.*

Student Book
pages 102–5

Place Study

You need to know:

* about the key differences between questionnaires and interviews
* how to analyse interview transcripts, using coding.

Advantages of interviewing

As a qualitative methodology, one-to-one interviews have a number of advantages:

* *Depth*: interviews provide an in-depth understanding of people's lives and their lived experience of a place, both in the past and present. In contrast, questionnaires push respondents into tick-box answers, which may not fit their actual experience.
* *Time*: they give people time to explain what they mean.
* *Flexibility*: interviews allow respondents to raise issues that the researcher may not have otherwise considered.
* *Additional sources*: some interviewees may recommend other people to talk to. They may also have useful personal photographs they are prepared to share. Such photographs can be used as a prompt, to help give your conversation focus and purpose (Figure **1**).

High Street is closed, as it is today for village events; flags also remain a feature of 'high days' in the village today.

The fire engine dates this image, featuring mid-century technology.

Proud family members and passers-by watch this community event.

Shop frontages have changed but the overall street-scene is still recognisable today.

⬆ **Figure 1** *An informal 'snap' of a parade along Great Missenden High Street to mark the coronation of Queen Elizabeth II (1953) provides an insight into lived experience of this place in the past*

Disadvantages of interviewing

One-to-one interviews also have disadvantages:

* *Time taken*: putting the interviewee at their ease will help the conversation to flow, but it takes time to build a rapport. In addition, getting access to people via local **gatekeepers** may take some effort; every interview will need to be set up at the interviewee's convenience.
* *Wider conclusions*: data generated will be rich but personal; meaning the wider conclusions you can draw from a handful of interviews may be more limited.

* *Sample size*: small – an, illustrative rather than a representative sample.
* *Subjectivity*: a common criticism; but arguably also true of questionnaire data. Either methodologies may generate subjective data.
* *Poorly designed questions*: so-called 'leading questions' are not the preserve of interviews alone. Whatever the technique used to talk to people, consider how you ask your questions. Are you suggesting the answer you would like to hear?

Continued over ≫≫≫

Place Study

Analysing and presenting your qualitative data

Given the wealth of the material you are likely to have after conducting (and recording) your interviews, you will need to be selective. Transcribe a few sections of your interviews in some detail, once you've listened again to the material you have collected.

Analysis: Dividing text up into themes and subthemes (or codes)

It can be useful to give each new theme within a conversation a title or **code** – make a list as you listen to your recordings. Making a note of the timings of specific codes or themes will make it easier to find relevant sections you want to focus on.

How to analyse transcripts:

1   Select 'changing employment' (as in Figure **2**) or another theme common to different interviewees.
2   Split any text about it into different subthemes (with sub-codes), e.g. comments that relate to particular employers or trades.
3   Subdivide your text further – into fact/opinion or experience/media knowledge.
4   Collate all of the interviewees' relevant comments that relate to a particular code (theme or subtheme). If you do this for a range of codes your data will be in an organised form ready for final review.
5   Consider, for instance, the consistency of what was said by individuals on a particular topic, or whether different respondents' accounts of changes in your local place dovetail or conflict. Can you explain these differences?

Final presentation: The power of a quote

Quotes (with commentary) can be usefully included in your independent investigation to:

- give a little local colour (a sense of place: the lived experience, the dialect)
- help to illustrate a point
- refute a finding that might otherwise be accepted as reliable or justified, given another set of (quantitative) data you have included in your study.

Part of the picture

Interviewing is a useful technique providing in-depth information, best used as part of a multi-method approach to a research question.

Valentine (2013) uses the analogy of 'triangulation', the use of different bearings to give a correct position, in her discussion of using oral sources alongside quantitative data to draw wider conclusions about phenomena in human geography.

Employment in the parish (E in)
- Agriculture (A)
- Manufacturing industry (M)
  - Electro-plating and enamelling – Gerhardi's (G)
    - Fact (fa)
      - Experience (exp)
      - Hearsay (hea)
      - Media knowledge (med)
    - Opinion (op)
  - Building trade – Wrights Yard (W)
- Service industry (S)
  - Shops and the High Street (H)
    - Grocers and home delivery service (D)
    - The Abbet book shop (A)

Employment out of the parish (E out)
- Local towns (T)
- London & commuters (L)

 Figure 2 *Themes and subthemes (or codes) may emerge from close analysis of interview transcripts – in this case, relating to employment in Great Missenden*

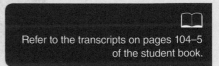
Refer to the transcripts on pages 104–5 of the student book.

⏱ **Sixty second summary**

- Interviewing, although time consuming, provides a wealth of data about people's lived experience of places and how they have changed over time.
- Coding is a way of categorising and collating sections of interview transcripts that relate to common themes (codes), and subthemes – it is a way of organising large amounts of qualitative data.
- Quotes, with commentary, could form a powerful element of the presentation of data within your independent investigation.

✏ **Over to you**

Can you define key terms that relate to the analysis of qualitative data (like provenance, reliable, subjective), at speed?

Student Book
pages 106–9

You need to know:

- about factors that have shaped the Detroit's social and economic geography.

Location and early development

Detroit's site and situation shaped its early development:

- The city is situated on the Detroit River between Lakes Huron and Erie, which connect to the Atlantic Ocean.
- Finding its location to be advantageous, French colonialists founded Fort Pontchartrain du Détroit in 1701, which became Detroit.
- Britain was another colonial master of the city in the late eighteenth century
- In the nineteenth century, shipping and shipbuilding brought wealth to the city.
- The so-called 'Gilded Age' mansions were built to the east and west of Detroit's downtown area; they demonstrated the economic rewards in this emerging transport hub.

Economic success and expansion

The Steel Belt boom

Detroit saw significant population growth in the nineteenth century, but it was in the twentieth century when its expansion (population and land area) really took off.

- Southern Michigan was part of the 'Steel Belt', where waterways and canals, roads and railroads helped to connect iron ore mines with coal resources.
- Home grown companies manufactured standardised products for customers in USA and, eventually, for export around the world, e.g. Ford and General Motors (Figure **2**).

New workers needed

During the twentieth century, millions of African Americans travelled from the rural southern states to the urban north-east, Midwest and west – an event termed the 'Great Migration'. This was driven by two linked *push factors*:

- Racism experienced by African Americans in the southern states. The white supremacist group, the Ku Klux Klan, had a great deal of support in the region and the authorities also officially favoured whites.
- The lack of economic opportunities for African Americans apart from labouring on plantations.

A distant, contrasting place

Sources of information about distant locations require critical appraisal. Which of the two apparently contrasting Detroits presented in this image, is the correct one? Or is the juxtaposition itself instructive about the lived experience of many residents in this place today?

▲ **Figure 1** *The Ford-financed Renaissance Center (on the left-hand side) was a failed attempt to regenerate Downtown Detroit*

▲ **Figure 2** *Worker on the assembly line at Chrysler's Jefferson North Assembly Plant, Detroit (2011)*

Continued over >>>

Growth of the city and the suburbs

Figure **3** shows the growth in population that took place during the twentieth century in Detroit – a significant *pull factor* for African Americans was the availability of new jobs in the automotive industry.

In the post-war period the population became increasingly dependent upon the car, and sprawled into a wider **metropolitan** area.

Competition and fuel insecurity

These suburbs, and their predominantly white, higher-paid residents, were not officially incorporated into Detroit. This was a crucial failure of the authorities, as tax receipts fell in the inner city area due to the crumbling industrial landscape (dubbed the **drosscape**), in the face of increased competition.

From the 1970s onwards, the economic tide was turning as international oil crises prompted drivers to buy vehicles with greater fuel economy. Competitors from Asia such as Honda, Datsun (Nissan) and Toyota produced more desirable models, causing a decline in sales and profits of Detroit's big employers (Ford, General Motors).

These companies linked to the automotive industry responded by cutting jobs and shutting down less efficient plants.

Milestones in Detroit's economic decline

- *In 2008*: Toyota became not only the leader in global production and sales in the auto industry, overtaking General Motors. In contrast, with economic decline in the late twentieth century, Detroit had become the capital of the **Rust Belt** (Figure **4**).
- *Also in 2008*: President Bush agreed a major financial bailout of Ford, General Motors and Chrysler.
- *In 2013*: the City of Detroit declared itself bankrupt – the largest municipal bankruptcy in the US.

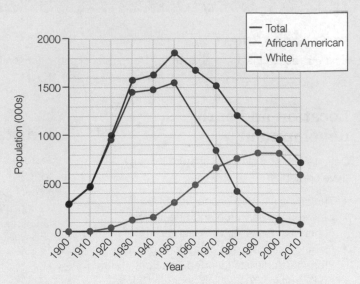

▲ *Figure 3* The growth in population in Detroit during the twentieth century (US Census)

▲ *Figure 4* The USA's successful Steel Belt became its Rust Belt in the late twentieth century

 Sixty second summary

- Detroit saw significant population growth fuelled by an economic boom.
- Detroit's population has been shaped by exogenous factors including the migration of African Americans from the poor working and living conditions of the southern states.
- From the mid-twentieth century onwards, the city's wealthier (predominantly white) population have moved out of the inner city to the metropolitan area's newer, outer suburbs, taking their taxes with them.
- Economic decline in the late twentieth century saw the City of Detroit and its major auto industry employers go bankrupt.
- International competition influenced growing job losses in the city, falling incomes and (indirectly) its landscape of industrial dereliction.

Over to you

Outline the endogenous and exogenous factors that have shaped Detroit, as it is today.

You need to know:

Student Book
pages 110–13

- about quantitative and qualitative sources that inform us about Detroit's changing demographic and cultural characteristics
- about a locality within Detroit: Alter Road, Grosse Pointe Park.

Demographic and cultural change in the twentieth century

Founded in 1959, the Motown record label become another successful brand of the city of Detroit. Motown fused the predominantly 'black' soul music with the predominantly 'white' pop sound. The company signed bands such as the Supremes, the Jackson 5 and the Spinners (known as the Detroit Spinners in the UK).

Mapping a segregated city

Your place study of Detroit should focus in on a particular locality and its local community, such as Grosse Pointe Park (Figure **1**). Could you draw a basic sketch map of your distant place?

Figure **2** is a racial dot map which shows the pattern of the predominantly African American population of the city, surrounded by the white population of Detroit's suburbs.

However, race riots in the 1940s and 1960s demonstrated that racial *harmony* was not always possible in the pursuit of the 'American Dream' in Detroit. In the 1970s, city authorities were accused of supporting the racial segregation of schools and housing, reinforcing **ghettos** and the racial divide for which the city is well-known today (Figure **2**).

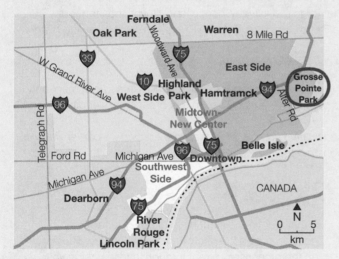

Figure 1 *Map of the districts of the city of Detroit and its surrounding suburbs*

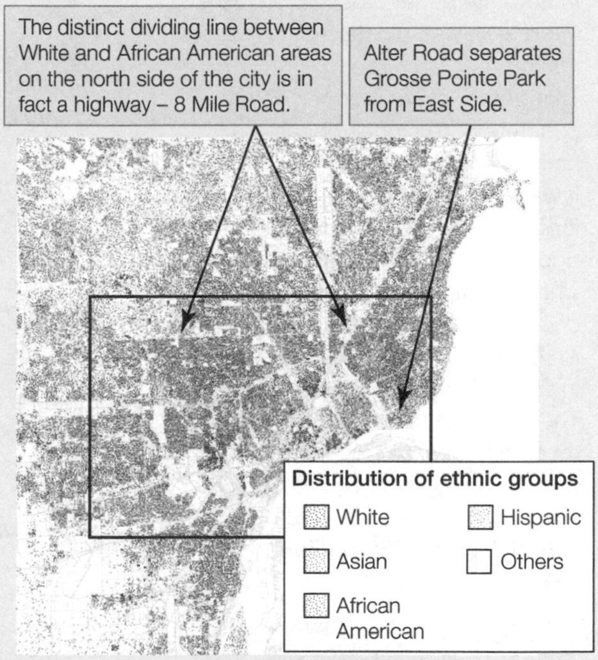

The distinct dividing line between White and African American areas on the north side of the city is in fact a highway – 8 Mile Road.

Alter Road separates Grosse Pointe Park from East Side.

Distribution of ethnic groups

- White
- Hispanic
- Asian
- Others
- African American

Figure 2 *A quantitative source: race and ethnicity map, Detroit (2010 US Census). Each dot represents 25 residents of the city, with different colours showing the distribution of different ethnic groups.*

Social processes create spatial patterns

White flight ...

In the second half of the twentieth century, Detroit's troubled times prompted white middle-class families to move out of the city to the suburbs of metropolitan Detroit. This migration was a choice made possible by their higher incomes, as they could afford to commute back into the centre.

Eventually, many jobs followed this relocation to the suburbs: a new socio-economic, spatial pattern was created by this process of 'white flight'.

... and racial integration?

In recent decades, a similar movement of African American middle-income families from the city to the suburbs has taken place. Though fewer in number, they also sought better schools, less crime and a higher quality of life.

Social commentators have debated whether or not this movement might reduce the level of segregation around, if not within, the city of Detroit.

Continued over >>>

Place Study

Place Study

Alter Road, Grosse Pointe Park: a national news story

In Detroit's affluent suburbs, which border the inner city area, there is contemporary evidence of the ways in which residents seek to maintain the divide between 'them and us', by creating new physical obstacles (Figure **3**).

In what appeared to be an official endorsement of the growing number of barriers between neighbourhoods in 2014, the authorities of Grosse Pointe Park undertook a range of measures to restrict the movement of traffic into the suburb from East Side (a district of the inner city).

- Kercheval Avenue, a major commercial thoroughfare that crosses Alter Road (Figure **1**), was blocked by a farmers' market.
- This barrier of wooden sheds and concrete curb stones became a national story.
- It was widely seen as a metaphor for the enduring division of Detroit's population; with the predominantly white, suburban middle classes divided from the predominantly African American, poorer residents who still live in Detroit's run-down inner city.
- There is a stark contrast between the median household income of the two populations: City of Detroit $26 955 (82% African American); Grosse Pointe Park $101 094 (85% white).

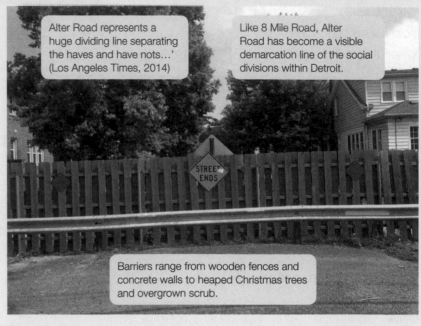

Alter Road represents a huge dividing line separating the haves and have nots...' (Los Angeles Times, 2014)

Like 8 Mile Road, Alter Road has become a visible demarcation line of the social divisions within Detroit.

Barriers range from wooden fences and concrete walls to heaped Christmas trees and overgrown scrub.

Figure 3 *A qualitative source: unofficial barriers block Alter Road frustrating drivers and pedestrians who attempt to cross from East Side, Detroit, into Grosse Pointe Park*

The (predominantly white) people of Grosse Pointe are likely to say that their relative wealth is the money that built Detroit. They mean it's old money, because it's where the descendants of the great auto-industry – particularly the Fords – still live. When, in fact, it was the labour of working-class Detroit that built Detroit. It's the lion's way of telling history.

(Social worker, Detroit)

Everyone is human, we all bleed red. Something like a 'racial map' only intensifies the division we already see in this great city. (Detroit resident)

Figure 4 *A qualitative source: blogposts may provide a partial view but give some insight into the lived experience in a place*

Sixty second summary

- Detroit's place-meaning has changed from being the home of mainstream black music, to a place of notoriety, known for race riots and violence on the streets.
- Alter Road is a modern-day dividing line between the poorer, predominantly African American neighbourhoods found in Detroit's inner city and the richer, predominately white suburbs, such as Grosse Pointe Park.
- The views of residents conflict about the need to recognise (and address) enduring racial segregation within the city.
- Quantitative and qualitative sources may be used together to understand changes in distant locations and get some sense of the lived experience in these localities.

Over to you

Whatever your distant place study, your approach to researching it is likely to have been quite different to the process you went through in researching your local place.

As you revise this place study, consider which sources were most useful to you. Which sources did you feel were reliable about a place you haven't visited? Which sources may require a more critical eye?

3
Contemporary urban environments

Your exam

(AL) *Contemporary urban environments* is an **optional topic**. You must answer **one** question in Section C of Paper 2: Human geography, from a choice of **three**: *Contemporary urban environments* **or** *Population and the environment* **or** *Resource security*.

Paper 2 carries 120 marks and makes up 40% of your A Level. Section C carries 48 marks.

(AS) *Contemporary urban environments* is an **optional topic**. You must answer **one** question in Section B of Paper 1: Physical geography and people and the environment, from a choice of **two**: *Hazards* **or** *Contemporary urban environments*.

Paper 1 carries 80 marks and makes up 50% of your AS Level. Section B carries 40 marks.

Specification subject content (specification reference in brackets)

Either tick these boxes as a record of your revision, **or** use them to identify your strengths and weaknesses

Section in Student Book and Revision Guide	1 🙁	2 😐	3 🙂	Key terms you need to understand Complete the key terms (not just the words in bold) as your revision progresses. 3.1 has been started for you.
Urbanisation *(3.2.3.1)*				
3.1 Global patterns of urbanisation				*growth, urbanisation, migration, urban expansion*
3.2 Forms of urbanisation				
3.3 Megacities and world cities				
3.4 World cities in global and regional economies				
3.5 Urban growth of Bengaluru				
3.6 Urban change				

Your revision checklist

3.7 GIS: planning and development			
3.8 Urban policy and regeneration in Britain since 1979			

Urban forms *(3.2.3.2)*

3.9 Urban form and characteristics			
3.10 'Movie-ing' to the city			
3.11 New urban landscapes			
3.12 The postmodern western city			

Social and economic issues associated with urbanisation *(3.2.3.3)*

3.13 Multiculturalism and cultural diversity			

Urban climate *(3.2.3.4)*

3.14 Urban microclimates			
3.15 Urban precipitation and wind			
3.16 Urban air pollution			

Urban drainage *(3.2.3.5)*			
3.17 Urban precipitation and drainage			
3.18 Drainage management			
3.19 River restoration and conservation			
Urban waste and its disposal *(3.2.3.6)*			
3.20 Urban waste and its disposal			
Other contemporary urban environmental issues *(3.2.3.7)*			
3.21 Other contemporary urban environmental issues			
Sustainable urban development *(3.2.3.8)*			
3.22 Sustainable urban development			
Case studies *(3.2.3.9)*			
3.23 Rio – an Olympic city			
3.24 London – an Olympic city			

You need to know:
- how important urbanisation is in human affairs
- the global patterns of urbanisation since 1945.

Student Book
pages 118–19

The importance of urban centres in human affairs

Cities influence our lives at many levels. They are important:

- for organising economic production, e.g. concentration of financial services
- the exchange of ideas and creative thinking, e.g. universities
- as social and cultural centres, e.g. theatres and national stadia
- as centres of political power and decision-making, e.g. government
- for the availability of labour, e.g. new migrants pulled into urban centres.

Big idea

- Urbanisation is the increase in the *proportion* of the population living in urban centres.
- Urban growth is the increase in the total population.
- Urban expansion is the increase in the physical size.

Traditionally, urbanisation has been associated with HDEs in the northern hemisphere.

Tokyo is still the largest city in the world.

Percentage of urban area population
- 81–100
- 61–80
- 41–60
- 21–40
- 0–20

City population (millions)
- 10≤ ●
- 5≤ ● <10
- 1≤ ● <5

a

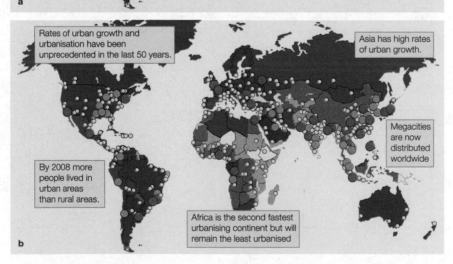

Rates of urban growth and urbanisation have been unprecedented in the last 50 years.

Asia has high rates of urban growth.

Megacities are now distributed worldwide

By 2008 more people lived in urban areas than rural areas.

Africa is the second fastest urbanising continent but will remain the least urbanised

b

Figure 1
Patterns of urban growth and urbanisation a) 1970 and b) 2014

Sixty second summary

- In 1945, less than a third of the world's population lived in urban areas. By 2030 it is expected that this will increase to two-thirds.
- Urbanisation has been a key process in all countries since 1945.
- Urban centres are important for human affairs such as exchanging ideas and decision-making.
- The number of large cities around the world has increased in rapidly in recent decades, particularly in LDEs.

Over to you

Write **three** bullet points on global patterns of urbanisation since 1945. Try to use Point, Evidence, Explain (P.E.E.) to structure each statement.

You need to know:

- about urbanisation, suburbanisation, counterurbanisation and urban resurgence.

Student Book
pages 120–1

The cycle of urbanisation

Processes of urbanisation (Figure **1**) usually take place at the same time in HDEs, although one process will dominate (Figure **2**). In LDEs, *urbanisation* is the main process but many Asian cities are beginning to show the effects of *suburbanisation* and even *counterurbanisation*.

See page 121 of the student book for a study of *urban resurgence* in Montreal.

1 Urbanisation: an increase in the proportion of a country's population living in a town or city.

4 Urban resurgence: population movement from rural back to urban areas. Associated with upwardly mobile young people, who are pulled to the centre of the '24-hour' city. This influx of youth and new wealth encourages a revival of some inner-city and CBD areas.

2 Suburbanisation: the **decentralisation** of people, employment and services towards the edges of an urban area. This outward growth, or **urban sprawl**, is closely linked to the development of transport networks.

3 Counterurbanisation: population movement from large urban areas to smaller urban settlements and rural areas. People move as a combined result of the push problems of the city (e.g. crime, congestion, land degradation) and the pull of rural life (e.g. bigger living space, 'safer' environment).

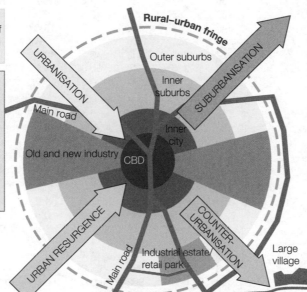

Figure 1 ▶

The four processes or cycle of urbanisation for a typical western European city

① **S & E England:** Counterurbanisation, associated with increased car ownership, has been dominant in the 1980s and 90s.

② **N & W England:** The 'economic pull' of cities in S & E England is a strong influence.

③ **Rest of UK:** Urban resurgence is now being experienced in many UK cities.

▲ **Figure 2** *Population change rate, by region, for three city sizes in the UK (1981–2010). Since the turn of the century (red column), cities of all sizes have shown growth.*

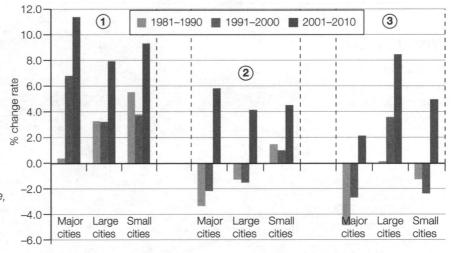

Sixty second summary

- Urbanisation is driven, primarily, by rural to urban migration but also by natural growth in youthful, urban populations.
- The cycle of urbanisation incorporates suburbanisation, counterurbanisation and urban resurgence.
- Counterurbanisation was a feature of UK cities in the 1980s and 90s.
- The population of all city sizes in the south and east of England increased between 1982 and 2010.

Over to you

Draw a simple sketch to summarise urbanisation, suburbanisation, counterurbanisation and urban resurgence.

3.3 Megacities and world cities

**You need to know:**

- about the emergence of megacities and world cities.

*Student Book*
pages 122–3

## Megacities

Megacities are city regions with a population of more than 10 million people.

Type of settlement		Total pop (millions)			% distribution			Growth rate (%)	
		1975	2000	2015	1975	2000	2015	1975–2000	2000–15
HDEs (growth rate slowing)	10 million or more	36	67	69	3.4	5.7	5.7	2.5	0.2
	5 million to 10 million	62	45	51	5.9	3.8	4.2	−1.3	0.8
	1 million to 5 million	145	219	250	13.9	18.5	20.6	1.6	0.9
	500 000 to 1 million	69	91	96	6.6	7.6	7.9	1.1	0.4
	Fewer than 500 000	422	481	503	40.2	40.5	41.4	0.5	0.3
	Rural areas	315	285	246	30.0	24.0	20.3	−0.4	−1.0
	Total population	1048	1188	1214	100.0	100.0	100.0	0.5	0.1
EMEs (urbanisation high for 1–5 million cities)	10 million or more	33	195	306	1.1	4.0	5.1	7.1	3.0
	5 million to 10 million	64	110	197	2.1	2.3	3.3	2.1	3.9
	1 million to 5 million	182	485	756	6.0	10.0	12.7	3.9	3.0
	500 000 to 1 million	106	209	277	3.5	4.3	4.7	2.7	1.9
	Fewer than 500 000	425	943	1314	14.0	19.4	22.1	3.2	2.2
	Rural areas	2217	2925	3091	73.2	60.1	52.0	1.1	0.4
	Total population	3026	4867	5940	100.0	100.0	100.0	1.9	1.3
LDEs (most still live in rural areas, but high growth rate in largest cities)	10 million or more	0	12	21	0.0	1.9	2.3	–	3.6
	5 million to 10 million	0	5	32	0.0	0.8	3.5	–	12.2
	1 million to 5 million	6	44	84	1.6	6.8	9.3	8.2	4.3
	500 000 to 1 million	6	15	17	1.7	2.3	1.9	3.8	1.0
	Fewer than 500 000	39	91	162	11.1	14.1	18.0	3.4	3.9
	Rural areas	298	477	586	85.6	74.0	64.9	1.9	1.4
	Total population	348	645	902	100.0	100.0	100.0	2.5	2.2

Figure 1 *The distribution by population size and development grouping*

World cities

World cities are the most important cities in the global economy.

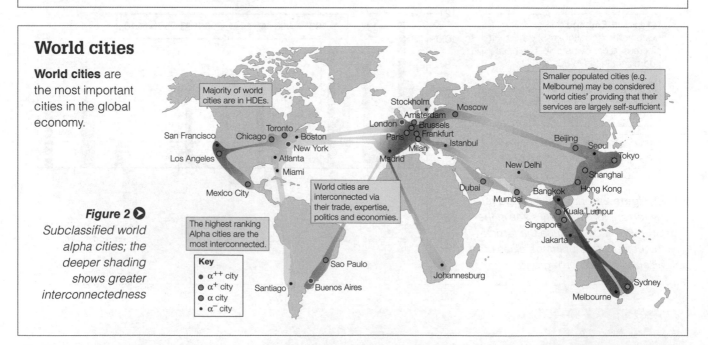

Majority of world cities are in HDEs.

Smaller populated cities (e.g. Melbourne) may be considered 'world cities' providing that their services are largely self-sufficient.

World cities are interconnected via their trade, expertise, politics and economies.

The highest ranking Alpha cities are the most interconnected.

Key
- α⁺⁺ city
- α⁺ city
- α city
- α⁻ city

Figure 2 ▶
Subclassified world alpha cities; the deeper shading shows greater interconnectedness

**Sixty second summary**

- Recent population growth has been most significant in LDEs and EMEs.
- 95% of future population growth in cities will be in LDEs and EMEs.
- World cities are linked together by trade, expertise and politics.

**Over to you**

Study Figure **1**. Analyse and estimate future trends in around **50** words.

74   **Chapter 3** – Contemporary urban environments

Student Book
pages 124–5

You need to know:

- the role of world cities in global and regional economies.

World cities and economic growth

World cities drive regional, national and global economies, support prosperity and create jobs.

The populations of most world cities are growing. However, counter-urbanisation is taking place in some HDEs, e.g. the population of New York is declining as the middle class leave.

Productivity tends to increase with city size, due to **agglomeration economies**.

But, this is not always the case. A lack of investment in cities in LDEs, especially in **infrastructure**, may result in slow or even negative economic growth.

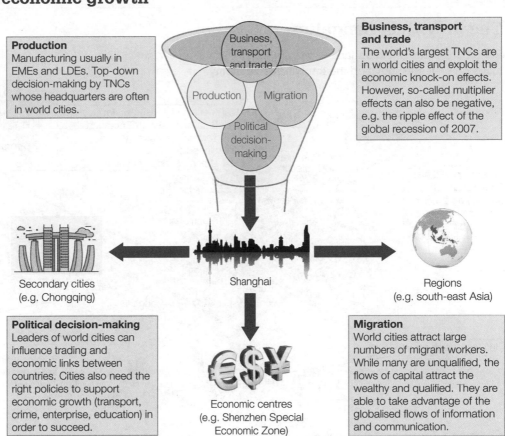

Production
Manufacturing usually in EMEs and LDEs. Top-down decision-making by TNCs whose headquarters are often in world cities.

Business, transport and trade
The world's largest TNCs are in world cities and exploit the economic knock-on effects. However, so-called multiplier effects can also be negative, e.g. the ripple effect of the global recession of 2007.

Secondary cities (e.g. Chongqing)

Shanghai

Regions (e.g. south-east Asia)

Political decision-making
Leaders of world cities can influence trading and economic links between countries. Cities also need the right policies to support economic growth (transport, crime, enterprise, education) in order to succeed.

Economic centres (e.g. Shenzhen Special Economic Zone)

Migration
World cities attract large numbers of migrant workers. While many are unqualified, the flows of capital attract the wealthy and qualified. They are able to take advantage of the globalised flows of information and communication.

Figure 1 *Interconnecting world cities act as funnels for economic growth; this growth flows to other cities that act as centres of further economic growth*

Shanghai

Shanghai is a world city on the east coast of China.
There are several factors that explain Shanghai's economic success:

- **Migration**: one quarter of the city's labour force has a college education.
- **Political decision-making**: as a result of its open city status and creation of development zones, there has been high (inward) investment.
- **Production**: new cities along the River Yangtze, support Shanghai's export-oriented economy.
- **Business**: favourable terms to foreign companies has resulted in disproportionate amount of overseas investment.

 Sixty second summary

- World cities act as funnels for economic growth, promoting growth in secondary cities and the wider region.
- Large cities in HDEs are more likely to become world cities than those in LDEs.
- Shanghai, on the east coast of China, is a world city that benefits from its large size as well as the productivity of new secondary cities along the Yangtze River.

 Over to you

Place-specific examples help to illustrate a point. With reference to Shanghai, draw a mind map to suggest reasons why world cities are so productive.

You need to know:
- the political, economic, technological, social and demographic processes associated with urbanisation and urban growth
- how Bengaluru has grown.

Student Book
pages 126–7

The growth of Bengaluru

Bengaluru is one of India's fastest growing cities. It is the equivalent of Silicon Valley focusing on technology and the **knowledge economy**.

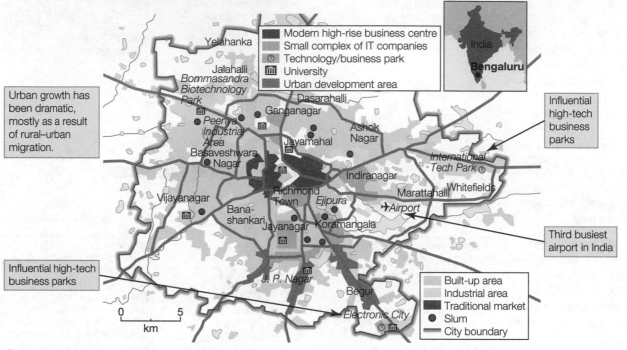

 Figure 1 *The growth of Bengaluru*

Process	Impact on urbanisation and urban growth of Bengaluru
Political	• Land set aside for high-tech business park – 'Electronic City'. • Low taxes, relaxed licensing laws, no limits on currency conversions. • Government investment tackles urban problems, e.g. housing, transport.
Economic	• 'Largest job creating city in India'. • Capital of aeronautical, automotive, biotechnology, electronics and defence industries. • Growth of informal sector to meet the needs of high-tech and corporate workforce.
Social	• **Trickle-down effects** of urbanisation have not benefitted all. • Growing middle classes live in gated communities in suburbs or in converted colonial properties nearer the centre. • Urban growth adding to pressure on housing, e.g. slums occupying marginal land.
Technological	• Colleges and universities provide highly skilled and (relatively) cheap workforce. • Overseas companies have been pulled into the city. • Growth of domestic IT based companies such as Infosys.
Demographic	• Rapid population growth; it is a megacity. • High in-migration rates, including from overseas. • Youthful age structure guarantees future continued population growth.

 Figure 2 *Processes impacting on urbanisation and the urban growth of Bengaluru*

Sixty second summary

- Bengaluru (Bangalore) is one of India's fastest growing cities.
- The city is the Indian equivalent of Silicon Valley with over ten million inhabitants and a focus on the knowledge economy.
- The state government set aside a large piece of land for a high-tech business park in the 1970s, an important political process.
- Economic, social, technological and demographic processes have also contributed to the rapid growth of Bengaluru.

Over to you

Rank ordering is an easy-to-use tool for prioritising. Rank order factors that helped Bengaluru to grow. Write down your reasons.

You need to know:

- about deindustrialisation, decentralisation and the rise of the service economy.

Student Book
pages 128–9

Deindustrialisation

This has taken place in HDEs since 1945 because of:

- the reduced need for labour – mechanisation and new technologies has led to increased productivity
- reduced demand for manufactured goods
- globalisation – outsourcing labour to LDEs and competition from EMEs
- increased costs of raw materials and those arising from subsidies and environmental controls.

Decentralisation

Businesses in HDEs move to peripheral locations because of:

- competition for space
- socio-economic and environmental *push* factors, e.g. retail/science parks
- *pull* of easily accessible outer suburbs and rural–urban fringe.

 Big idea

- **Deindustrialisation**: the long-term decline of a country's industries.
- **Decentralisation**: the movement of people, businesses/functions to settlements from the centre to the periphery.

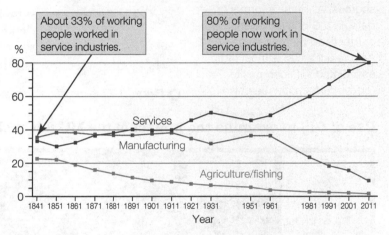

The UK's structural shift in employment (1841–2011). The falling green and blue lines show the effects of deindustrialisation and shift to a post-industrial economy.

Rise of service economy

This shift in employment in HDEs since the 1950s has been caused by:

- growth in corporate (multinational) headquarters
- rise of knowledge economy to meet the needs of other businesses
- growth in research and development
- expanding consumer demand for leisure services
- property development
- an increase in tourism.

However, this shift has not benefitted all:

- well paid managerial jobs are few in number
- 'back office jobs' are lower paid and often temporary
- more manufacturing jobs lost than service jobs gained
- urban unemployment may remain high.

Large multinationals in the largest cities attract services. Consumer services increasingly favour locations near affluent suburban populations. Research and development facilities are attracted to urban fringe locations near universities.

 Sixty second summary

- Cities are the result of interactions between the people living there.
- Deindustrialisation is the long-term decline of a country's manufacturing and heavy industry.
- Deindustrialisation is caused, in part, by reduced demand, increased costs and the globalisation of manufacturing.
- UK employment in agriculture/fishing, and manufacturing have fallen since 1841.
- The twentieth century saw a dramatic rise in the UK's service economy.

 Over to you

Make sure that you can clearly explain the differences between deindustrialisation and decentralisation.

You need to know:
- the role that GIS plays in urban planning and redevelopment
- how GIS is applied.

Student Book
pages 130–1

What is GIS?

A Geographical Information System (GIS) is a computer database system capable of capturing, storing, analysing and displaying geographical information from a wide variety of sources. All of the information, or data, in a GIS is georeferenced – or identified by its location. As a result a GIS is most often associated with composite (or overlaid) maps (see also 2.8).

1905	Herbert Austin forms Austin Motor Company Limited at Longbridge.
1914–19	Rapid expansion of site associated with war effort.
1968	Austin eventually amalgamated into British Leyland.
1975	British Leyland refinanced by government.
2005	MG Rover group collapses.
2007/08	Chinese companies, Nanjing Automotive and then Shanghai Automotive Corporation form MG Motor UK. Research and development moves to Longbridge.
2011	New MG rolls off production line
2016	Car production stops; all MG vehicles now imported from China.

⌃ **Figure 1** *Timeline of car industry in Longbridge*

Use of GIS to examine the impact of the MG Rover closure, Longbridge

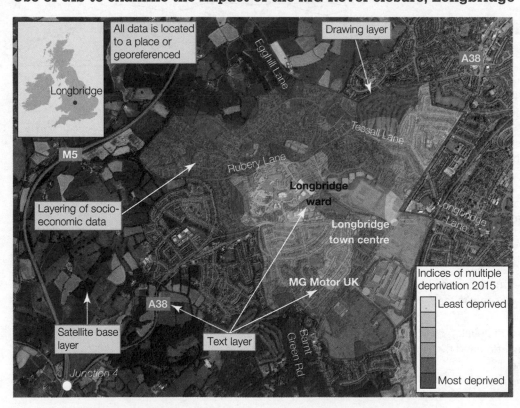

◁ **Figure 2** *Longbridge ward, including indices of deprivation*

 Sixty second summary

- Advances in ICT (Information and communication technology) have enabled geographers to use databases and present them in visual form using maps.
- GIS gives geographers the ability to use text, figures and tables with satellite images and maps.
- GIS, which commonly uses composite maps (or overlays), makes it possible to integrate information that is difficult through any other means.
- The changing fortunes at the MG Rover plant in Longbridge has had a direct impact on levels of **deprivation** in the local community.

 Over to you

In your examination, you are likely to have to think like a GIS – in other words, to work with and analyse layers of geographical information.

Write down **three** impacts of the closure of MG Rover.

You need to know:

- about urban policy and regeneration in Britain since 1979.

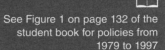
Student Book
pages 132–3

Urban regeneration in UK

The years between 1979 and 1990 saw the biggest shake-up in urban policy and **regeneration** in the last 50 years. Successive governments have continued to implement change, promote growth and development in our urban areas.

See Figure 1 on page 132 of the student book for policies from 1979 to 1997

Phase	Year	Policy	Comment
1997–2010 Labour government. Many of the schemes initiated by the Conservative government continue, albeit in another guise. Increasing number of **QUANGO** initiatives and a move towards more sustainable communities.	1997	Single Regeneration Budget (Challenge Fund)	Emphasis on local priorities, housing, crime and unemployment particularly targeted (revised May 2001).
	1997	Regional Development Agencies	Coordination of regional economic development and regeneration. Wide-ranging powers.
	1999	English Partnerships (EPs)	Responsible for land acquisition and major (sustainable) development projects, alone or in private partnerships.
	1999	Urban Regeneration Companies (URCs)	Organised and delivered major regeneration projects in key urban locations.
	A wave of government and QUANGO initiatives emerged at the end of the twentieth century, including Education Action Zones, Health Action Zones and Millennium schemes.		
2010–15 Coalition government adopts a decentralised and localist policy agenda.	2010	Local Enterprise Partnerships (LEPs)	Replaced RDAs whose functions largely passed to central government.
	2010	New Homes Bonus	Incentive to local authorities to accept housing growth (councils receive double the council tax per home for six years).
	2010	Community Infrastructure Levy	Neighbourhoods receive a proportion of the funds that councils raise from developers.
	2011	Localism Act and Tax Increment Funding (TIF)	New powers for city leaders to make local policies and promote economic growth.
2015–present Conservative government pushes for corporate social responsibility – or the benefits of regeneration to the whole community.	2016	Northern Powerhouse, Midlands Engine, Northern Ireland regeneration projects	Wide-ranging and large-scale regeneration projects to tackle regional (economic) differences. Aim to attract overseas inward investment.

Figure 1 *Urban regeneration schemes 1997–2017*

The regeneration of Longbridge

The transformation of the Longbridge site is the largest regeneration scheme outside London. By 2035 it should have created 10 000 new jobs and built 2000 new homes and has already achieved:

- £66 million Bournville College
- £8 million internet Connectivity Package
- £100 million Longbridge Technology Park

- £23 million Cofton Centre Industrial Park
- Austin Park, a 3-acre green site
- 19 000m² of new office and industrial space
- The Factory – a flagship youth centre
- 350 new homes within walking distance of the rail station
- The £70 million Longbridge 'town centre' is a mixed use retail, leisure, educational and residential development.

Sixty second summary

- The Labour government introduced Regional Development Agencies and English Partnerships from 1997.
- From 2010, the Coalition government's urban policy included a New Homes Bonus and Local Enterprise Partnerships.
- The Conservative government shifted focus towards large-scale regional regeneration projects with emphasis on increased corporate social responsibility.
- Several methods of regeneration have been used in Longbridge following the closure of the car plant.

Over to you

When used to support a statement, selective use of figures is to be encouraged. How many numbers do you remember from this page? Which makes the biggest statement?

Write down **three** to **five** points on urban regeneration of Longbridge, supporting **each one** with numerical evidence.

Student Book
pages 134–7

What influences urban form?

Urban form changes in response to:

- population – as populations become more mobile, flows of people are increasingly difficult to predict and to manage
- environment – e.g. physical infrastructure need to keep pace with population change
- economy – e.g. industries in HDEs follow population to locations beyond the urban-fringe
- technology – e.g. **teleworking** is challenging the notion of work hubs in the city
- policies – e.g. government approach towards housing, planning, transport and the economy, restrict or encourage urban change
- 'shocks' – e.g. economic recession, mass migration, extreme weather events, resource insecurity and demographic changes.

 Big idea

Urban form is the physical characteristics of built-up areas including the shape, size, density and make-up of settlements.

Contemporary characteristics of megacities

Weak planning systems combined with rapid population growth mean that megacities may have a diversity of urban form. Nevertheless most megacities share some or all of the following characteristics:

- Urban sprawl and peripheral growth
- **Edge cities**
- High density living and intensification of urban centres
- **Residential differentiation** (different social groups live apart)
- Redevelopment and conservation
- Ageing infrastructure
- Transit-oriented development
- Car-dominated urban form
- Environmental problems

What is infrastructure?

Infrastructure is the physical structures needed for society to operate (Figure 1). It includes energy, transport, water, waste water, solid waste, ICT, cultural, social, green and blue infrastructure (interconnected areas of land and water).

Megacities may struggle to provide infrastructure for increasing numbers of people. This is because for example, power stations, sewer systems and transport networks may be largely fixed (or at least require huge investment if they are to change and/or adapt). Furthermore, mixed ownership means that infrastructure may be slow to respond to changes in urban form.

▲ **Figure 1** *Bus, train and cable car stations of Stalden, near Zermatt. Switzerland is world-renowned for building large-scale infrastructure projects, often in challenging mountainous environments.*

Comparing and contrasting different urban forms

Characteristic urban form	Reasons for urban form	Functional zones	Example
Pre-industrial cities	Largely unaffected by industrial developments and have retained much of their urban layout and characteristics. Elite groups tended to locate in the centre surrounded by the lower socio-economic groups.	Historic buildings dominate the centre. High-class residential zone(s) near centre. Delineation of residential and commercial districts less clear than today.	Lincoln & Bath, UK Carcassone, France
Modern (or industrial) cities	Similar activities and similar people group together. Arrangement of areas determined by the general decline in land values outwards from the city centre.	Dominant CBD. Residential zoning. Industrial zone likely to be manufacturing-based.	Birmingham, UK Chicago, USA
Post-industrial cities	Urban mosaic – more chaotic and looser structure with many smaller zones rather than one or two dominating.	Multi-nodal structure. Less dominant CBD. Higher degree of social polarisation. Service sector-based industry less tied to one location.	Milton Keynes, UK Tokyo, Japan Las Vegas, USA
Public transport oriented (PTO) cities and motor based cities (MBC)	PTO development takes an integrated approach towards planning, e.g. short distances between housing and public transport nodes. MBC – mass motorisation from the 1950s increased suburbanisation rates and decentralisation of some economic activities.	PTO cities may develop along railway lines and main roads. MBC development linked to major road networks – urban freeways or motorways. Non-residential land uses, such as retailing and offices may locate in urban fringe locations.	Hong Kong (PTO) Detroit (MBC), USA
African cities	Many cities have grown from colonial settlements and not experienced industrialisation. Recent and rapid urban growth forced changes to established, older zones as well as peripheral expansion. Lacking resources and control, urban form may be unplanned and chaotic.	Dominant CBD likely to be the political, cultural and historic centre. Older industrial areas adjacent to transport routes. Newer peripheral middle-class housing served by road network. Informal housing developments on marginal land.	Nairobi, Kenya Cape Town, South Africa
Socialist cities	Classless city principle – everyone should live in same type of housing block irrespective of the location in the city. Housing blocks located close to local services. City centres administrative and political rather than commercial.	Homogenous blocks throughout the city. Neighbourhoods had local services; districts had higher order shops and entertainment services. City centre had prestige buildings and a central square for socialist rallies.	Prague, Czech Rep

 Sixty second summary

- Urban form is the physical characteristics of built-up areas, including shape, size, density and configuration of settlements.
- Urban form changes over time in response to factors such as the economy, population and technology.
- Most megacities share similar characteristics, including peripheral growth, edge cities, infrastructure and high-density living.
- There are several different types of characteristic urban form, each with their own typical range and layout of different functional zones.
- Cities continue to be shaped by processes of urbanisation, suburbanisation and counterurbanisation.

 Over to you

There are many characteristics of megacities and a wide range of different urban forms. Remembering them all is difficult. Create a mnemonic (or a system of patterns of letters or associations) to help you remember.

You need to know:

- The characteristics of Los Angeles and Mumbai.

Student Book
pages 138–9

A comparison of Los Angeles and Mumbai

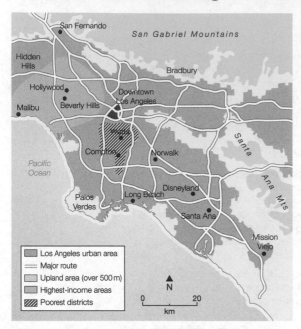

Los Angeles urban area
Major route
Upland area (over 500 m)
Highest-income areas
Poorest districts

⊙ Figure 1 *Greater Los Angeles (LA), on the west coast of California, USA*

⊙ Figure 2 *Mumbai situated on the west coast of Maharashtra, India*

	Los Angeles	Mumbai
Location	Greater Los Angeles has a population of close to 13 million and sprawls for 115 km east to west and covers 1166 km².	Mumbai squeezes more than 12 million people into an area of just 438 km².
Land use	• Largely white, car-owning, middle classes moved to LA to escape problems of industrial cities. • Increasing car ownership has led to suburbanisation and subsequent urban sprawl. • Low density housing including 88 'cities'.	• One of most crowded cities in the world. • Skyscrapers dominate Fort area (CBD). • Luxury high-class enclaves (e.g. Malabar Hill) live in stark contrast to half of the population who live on the streets or in slums (bustees).
Economic inequality	• The centre ('Donut city') dominated by headquarters of TNCs that offer few jobs for lower skilled. • Employers offering lower-paid service jobs moved to **exurbs** or edge cities (e.g. Anaheim). • Exclusive high-class 'cities' (e.g. Beverly Hills) border deprived areas which suffer from unemployment, high crime rates and poor access to schools and healthcare.	• India's commercial, economic, transportation and cultural centre attracts inward investment (partly because of skilled workers and low wages). • Small number of wealthy; majority live in poverty and service a large informal economy. • Lack of investment in urban infrastructure leaves millions at risk (e.g. drinking untreated water).
Cultural diversity	• Multicultural city. • Mexican Hispanics are largest ethnic group. • Ethnic enclaves (e.g. Chinatown, Little Italy and Watts district) are partly a result of 'white flight' to the suburbs and may suffer from multiple aspects of deprivation.	• 1000 new migrants per day, adding to cultural diversity. • Many faiths practised, including Buddhism, Islam, Hinduism and Christianity. • Several languages spoken, including 'Bambaiya Hindi' and Marathi (the mother tongue of local Maharashtrians).

Sixty second summary

- Los Angeles is geographically a huge city which typifies urban sprawl.
- The centre of LA has few opportunities for unskilled workers.
- There are many ethnic enclaves in the LA Downtown (inner city) area.
- Mumbai is one of the world's most crowded cities putting pressure on its urban infrastructure.
- Despite a growing middle class, Mumbai is still a city of huge contrasts.
- Many live in poverty and approximately 1000 migrants arrive each day.

Over to you

Map annotation is a way to remember detail of named places. Sketch a map of Los Angeles and annotate characteristics of land use, economic inequality and cultural diversity.

Student Book
pages 140–1

You need to know:

- about town centre mixed-use developments, cultural and heritage quarters, fortress developments, gentrified areas and edge cities.

Mixed-use development

Town centre mixed-use developments include:

- interconnected residential, commercial, cultural, institutional and industrial uses
- multiple uses that are safely and easily accessed by pedestrians
- Longbridge Village (see 3.8) and BedZed (3.22).

Cultural and heritage quarters

Many UK cities, such as Cardiff, have developed a cultural quarter to encourage growth and revitalise the local economy in the arts and creative industries.

Fortress developments

Fortress developments use walls, guarded entrances, security firms and CCTV to defend or protect areas or spaces. Such developments are increasingly found in urban areas, e.g. gated residential developments and shopping centres. Partly because of the costs involved, these developments are usually used by higher socio-economic groups, and therefore tend to promote social segregation.

Gentrified areas

Gentrification is a form of inner-city regeneration. It usually involves:

- affluent middle-class people move to traditionally run-down, cheaper inner city areas
- property price increase of gentrified homes and adjacent areas
- wider multiplier effects – local services upgrade (e.g. pubs to wine bars).
- the exclusion of the less affluent and established local population
- an affordable route onto the property ladder.

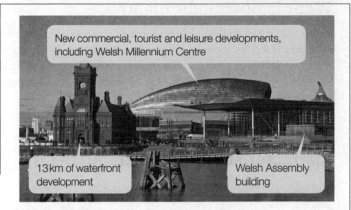

New commercial, tourist and leisure developments, including Welsh Millennium Centre

13km of waterfront development

Welsh Assembly building

⬆ **Figure 1** *The creation of Cardiff Bay facilitated a new quayside cultural and heritage quarter*

Edge cities

Edge cities are a consequence of suburbanisation and decentralisation. They are characterised by mixed office, residential and leisure spaces and tend to be located in the outer suburbs, near to motorway or main road junctions.

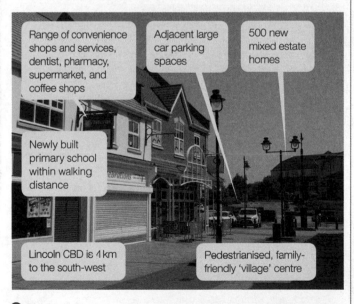

Range of convenience shops and services, dentist, pharmacy, supermarket, and coffee shops

Adjacent large car parking spaces

500 new mixed estate homes

Newly built primary school within walking distance

Lincoln CBD is 4km to the south-west

Pedestrianised, family-friendly 'village' centre

⬆ **Figure 2** *Bunkers Hill, Lincoln – a mixed edge development*

 Sixty second summary

- Town centre mixed-use developments blend different aspects of interconnected urban landscape.
- Many cities now have a heritage or cultural quarter, which helps to revitalise the arts and creative industries.
- Fortress developments do not encourage social mixing.
- Gentrification is a form of inner-city regeneration.
- Edge cities are suburbs that have developed into city-like centres.

Over to you

Write down a concise glossary for the following new urban landscapes: town centre mixed-use developments, cultural and heritage quarters, fortress developments, gentrified areas and edge cities.

You need to know:

- about characteristics of the postmodern western city.

Student Book
pages 142–3

What is postmodernism?

Whilst difficult to define, postmodernists see western world society as an outdated lifestyle that is impersonal and faceless. They embrace a variety of lifestyles and points of view, for example, celebrating ethnic diversity. Postmodern cities incorporate different, and sometimes spectacular forms of architecture (Figure **2**).

Big idea

A postmodern city reflects the changed social and economic conditions of the late twentieth century in some western cities.

Characteristics of a postmodern city

Change associated with urban form	Characteristic
Urban structure	• Chaotic multimodal structure • High-tech corridors • Post-suburban developments
Urban architecture and landscape	• Unusual mix of different styles and shapes that incorporate meaning and symbolism • Celebration of the past through historical references
Urban government	• Encouragement of mobile international capital • Services provided by the market (rather than public services) • Public and private sectors work in partnership
Urban economy	• Service (and especially quaternary) sector dominated • Globalised • Consumption oriented
Planning	• Spatial 'fragments' designed for aesthetic rather than social ends • Attempts to incorporate the views of many stakeholders
Culture and race	• High levels of social polarisation • Highly fragmented

 Figure 1 *Characteristics of the postmodern city*

Las Vegas – a postmodern city

Gaming and casino industry is a powerful lobby group that dominates a weaker local government.

Postmodern urban architecture draws inspiration from global landmarks.

Ecology of fear – Nevada has a high crime rate.

Tens of thousands of tourists walk the Strip promenade each day.

24-hour consumerism, particularly tourism, drives the economy.

The four mile 'Strip' is an eclectic mix of styles—this is partly because builders are allowed to build where and what they want.

Lake Las Vegas is a 30 minute drive into the Nevada desert. This post-suburban gated development includes mock Mediterranean villas on the shores of a human-made 'Italian lake'.

 Figure 2 *Las Vegas – an example of a postmodern city*

Sixty second summary

- Many western cities can be seen as examples of **postmodernism** due to changes that have taken place there.
- Changes associated with urban form in a postmodern city include those in urban structure, architecture, government, economy, planning and culture and race.
- Las Vegas displays all the characteristics of a postmodern western city.
- Changes in Las Vegas include over-the-top architecture, an economy dominated by tourism and a planning system giving builders freedom to build what and where they want.

Over to you

You need to understand key terms and use them in context. Write down the meaning of the following terms – postmodern city, post-suburban development, postmodern architecture.

Student Book
pages 144–7

You need to know:

- what multiculturalism is
- the issues associated with, and strategies to manage, cultural diversity, economic inequality and social segregation.

What is multiculturalism?

Culture is loosely defined as the way of life, especially the customs, language and beliefs (religion), of a particular group of people.

Multiculturalism is often a sensitive and emotive issue with a range of meanings. One interpretation is that every culture has the right to exist and might benefit from exposure to other cultures; and aspire to happy coexistence. Britain has a long history of immigration and is an example of a multicultural society.

Issues associated with cultural diversity

- *Economic* – migrants may meet labour shortages particularly in the service (e.g. the NHS) and manufacturing sectors. But, the perception of 'jobs for migrants' may cause resentment and racial intolerance.
- *Housing* – multiple occupancy in rented accommodation, usually within the poorer inner city, is widespread. *Residential succession* is when established ethnic groups move to suburban locations leaving inner-city housing available for newly arrived migrant groups.
- *Education* – schools may become dominated by one ethnic group; this impacts on the curriculum – additional English lessons may be necessary and special religious provision may be requested by parents. There may also be significant variation in educational attainment of different ethnic groups.
- *Health* – many ethnic minority groups live in inner-city areas, where there tends to be poorer levels of health. However, this is more a result of the poorer quality of the built environment than the underlying health of the population.
- *Religion* – following religious calendars can lead to friction with employers and local communities. Religious or cultural practices, such as wearing traditional clothing, can also be a potential cause of conflict.
- *Gender* – women from ethnic groups may require more flexible working as they are often the primary carers of both old and young family dependents – part-time jobs are often poorly paid.
- *Food culture, pop music and sport* – all help to support a blending of cultures (interculturalism – Figure **2**).
- *Language* – not knowing the host country's language will act as a barrier to integration and restrict employment and educational opportunities.

◀ **Figure 1** *In 2005, Royal Mail issued a set of stamps showing, by means of ethnic stereotypes, 'the diversity of British cuisine in today's multicultural society'*

▲ **Figure 2** *A multicultural squad of players helped Leicester City win the English Premier League in 2015–16.*

Continued over ▶▶▶

Cultural diversity in Batley, West Yorkshire

Batley is an example of a culturally diverse society.

- *The number of ethnic minority groups* has increased in Batley and Kirklees in recent decades.
- *Ethnic diversity within families* – the proportion of households with two or more people whose birth parents are from different ethnic groups has also grown (Figure **3**).
- *Language* – Gujarati, Punjabi and Urdu are widely spoken; a challenge for local government and schools.
- *Education* – around one in three schools in Kirklees is a faith school.
- *Poverty* – Batley is a deprived community and suffers from multiple aspects of deprivation. However, poverty is largely a result of the underlying difficulties with the built environment.
- *Community action* – is a strong feature of the town (e.g. the Indian Muslim Welfare Society works with individuals of all faiths).
- *Housing* – the Sadeh Lok Housing Group is one social housing provider that builds new and affordable homes.

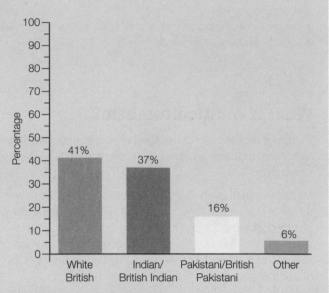

Figure 3 *Ethnic diversity in Batley East*

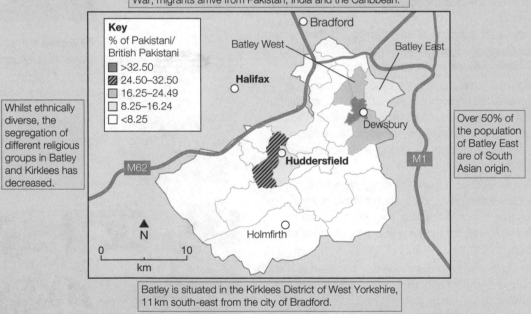

Figure 4 *Batley, near Bradford, is an example of cultural diversity*

Sixty second summary

- The UK has a long history of immigration; London is one of the world's most multicultural cities.
- Immigration into the UK remains a controversial, political topic.
- There are many issues associated with increased cultural diversity, including impacts on the economy, schools, healthcare and food.
- Batley is culturally diverse and suffers from multiple aspects of deprivation. However, poverty is a result of the built environment rather than being a particular feature of minority ethnic groups.

Over to you

The command word *discuss* demands a case *for* and *against* an argument, usually with examples used in support. Write down **three** points *for*, and **three** points *against*, an increasingly multicultural Britain.

Student Book
pages 148–9

You need to know:

* about the impact of urban forms and processes on local climate and weather
* about urban temperatures and the heat island effect.

Urban microclimates

Cities create their own **microclimates** – effectively *climate domes* with their own distinctive:

* temperature ranges
* humidity, precipitation generation and patterns
* wind speeds, turbulence and eddies.

Figure **1** shows how urban structure and activities create this climate dome.

Within this climate dome there are two levels – an urban canopy below roof level (where processes act in the spaces between buildings) and the urban boundary layer above roof level. Prevailing winds extend these urban patterns of precipitation and air quality to the immediate rural area downwind.

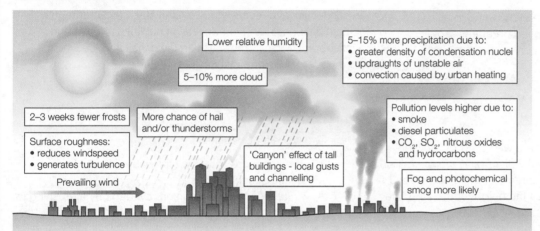

Lower relative humidity

5–10% more cloud

5–15% more precipitation due to:
• greater density of condensation nuclei
• updraughts of unstable air
• convection caused by urban heating

2–3 weeks fewer frosts

More chance of hail and/or thunderstorms

Pollution levels higher due to:
• smoke
• diesel particulates
• CO_2, SO_2, nitrous oxides and hydrocarbons

Surface roughness:
• reduces windspeed
• generates turbulence

'Canyon' effect of tall buildings - local gusts and channelling

Fog and photochemical smog more likely

Prevailing wind

◀ Figure 1
Characteristics of urban microclimates

The urban heat island

The best known microclimatic impact of built-up areas is the *urban heat island* effect. The urban 'island' is significantly warmer than the surrounding rural 'sea' due to:

* lower **albedo** – extensive dark surfaces (e.g. tarmac) absorbing heat
* large expanses of glass and steel reflecting heat
* less vegetation (evapotranspiration) and surface water to create humidity
* poor building insulation
* urban areas generate their own heat (industry, vehicles, inhabitants).

The temperature decline from urban centre to rural–urban fringe is known as the *thermal gradient*. This difference can be more than 6 °C in late summer and around 2 °C in winter.

Figure 2 ▶
Temperature distribution across London (August 2013)

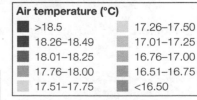

Air temperature (°C)

■ >18.5	17.26–17.50
■ 18.26–18.49	17.01–17.25
■ 18.01–18.25	16.76–17.00
■ 17.76–18.00	■ 16.51–16.75
17.51–17.75	■ <16.50

Sixty second summary

* Urban areas can have their own temperatures ranges, precipitation patterns, humidity, wind speeds and visibility levels.
* Factors contributing to the urban microclimate include pollution levels from vehicles and factories, along with tall buildings channeling winds.
* An urban heat island describes a built-up area where temperatures are significantly warmer than the rural areas that surround it.

Over to you

Produce a spider diagram that summarises the characteristics of urban microclimates.

You need to know:

- how urban environments affect precipitation and wind.

Student Book
pages 150–1

Urban precipitation

For precipitation to form, air has to rise and cool below the **dew point** – turbulence and heat island-related convection promote this uplift. Dust and pollution increases the density of condensation nuclei and 'seed' the cloud droplets – there is between 5% and 15% more rainfall in urban areas (Figure **1**, 3.14).

Snow melts relatively rapidly owing to generally higher urban temperatures and heat-retaining darker road and roof surfaces.

Fog

Fog is effectively cloud at ground level. Again, the higher concentration of condensation nuclei over cities encourages their formation. Just as Britain experienced more days of winter fog before clean air legislation, Beijing and New Delhi suffer similarly today – caused by vehicle emissions and coal burning.

Thunderstorms

Urban convection is especially powerful in summer. Updraughts of hot, humid air (including condensation nuclei) rise, cool and condense rapidly. This forms water droplets, hail and ice which charge thunderclouds and discharge as lightning.

Wind

Urban structures interfere with wind by slowing, redirecting and generally disturbing the overall airflow. As a general rule, the effects of buildings extend downwind by ten-times the height of the structure (Figure **1**).

Two key effects result:

- *urban canyons* – narrow streets between high-rise buildings funnel winds
- *venturi effect* – gusting in narrow gaps by air rushing to replace low pressure in the lee of buildings.

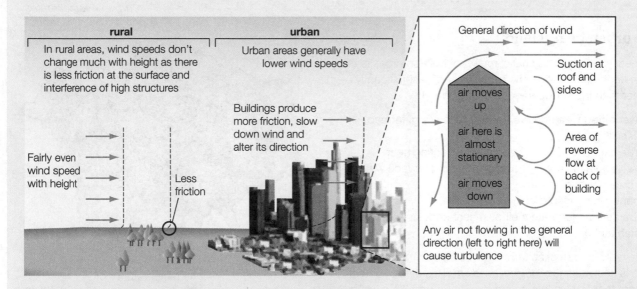

rural	urban
In rural areas, wind speeds don't change much with height as there is less friction at the surface and interference of high structures	Urban areas generally have lower wind speeds

Fairly even wind speed with height

Less friction

Buildings produce more friction, slow down wind and alter its direction

General direction of wind

Suction at roof and sides

air moves up

air here is almost stationary

air moves down

Area of reverse flow at back of building

Any air not flowing in the general direction (left to right here) will cause turbulence

 Figure 1 *Tall buildings interfere with airflow in urban areas*

Sixty second summary

- Urban areas have a higher rate of precipitation than rural areas, owing to warmer temperatures and higher levels of pollution.
- The higher concentration of condensation nuclei over cities also encourages the formation of fog and mist.
- Urban convection leads to intense summer thunderstorms.
- Urban canyons in cities can interfere with airflow resulting in the venturi effect.

Over to you

Produce a series of flashcards to record information about each of the microclimatic impacts on this and the previous page (urban heat island effect, precipitation, fog, thunderstorms and wind).

Student Book
pages 152–3

You need to know:

- about air quality: particulate and photochemical pollution
- how pollution can be reduced.

Urban air quality

Poor urban air quality is associated with:

- carbon dioxide enhancing the greenhouse effect
- nitrous oxides and sulphur dioxides causing haze, respiratory problems and acid rain
- carbon monoxide causing heart problems and tiredness
- *photochemical* oxidants causing eye irritation and headaches
- **particulates** causing respiratory problems and dirty buildings.

Smog

Smog occurs when smoke particulates and sulphur dioxide from burning coal mix with fog:

- The infamous 1940–50s London smogs were caused by sinking cold air during anticyclones trapping pollutants in a pollution dome.
- Smog created by temperature inversion is more commonly associated with cold air sinking to the floor of sheltered valleys and displacing warmer air above.
- Today, the cause of smog is likely to be photochemical – sunlight reacting chemically with industrial and vehicle emissions to form a cocktail of secondary gases (Figure 1).

⊙ **Figure 1** *Photochemical smog – Los Angeles, California, USA*

Pollution reduction

Reducing air pollution is a major challenge, particularly in rapidly developing LDEs and EMEs. Strategies involve a mixture of technical innovations, legislation and vehicle restrictions (Figure 2).

Development of hybrid electric (HEVs), 'plug-in' hybids (PHEVs) and fully electric vehicles (EVs)

Lean-burn engines, lead-free petrol and catalytic converters on vehicle exhausts

Clean Air Acts throughout HDEs

Complete ban of all except emergency, utility and electric vehicles

City centre pedestrianisation and Park and Ride schemes

Congestion charging

Legislation

Technical innovations

Vehicle restrictions

Strict controls on emissions, including smoke-free zones

Reduced industrial gas and particulate emissions using filters

Air quality targets and pollution warnings

Selective vehicle bans determined on specified days

 Figure 2 *Strategies to reduce air pollution*

 Sixty second summary

- Air quality in urban areas is worse than that in rural areas.
- Smog (**sm**oke and f**og**) was historically caused by temperature inversion trapping pollutants.
- Photochemical smog is caused by sunlight reacting with industrial and vehicle emissions.
- Strategies to reduce urban pollution include legislation on emissions, restrictions on vehicles and technical innovations.

Over to you

Why might urban pollution be reduced in the future in HDEs but increase in LDEs?

You need to know:

- about urban precipitation, surfaces and catchment (drainage basin) characteristics
- the impacts on drainage basin storage areas
- about the urban water cycle.

Student Book
pages 154–5

Urban precipitation, surfaces and catchment characteristics

Urban areas have more precipitation than rural areas because:

- warmer air holds more moisture
- pollution makes more condensation nuclei.

However, there is less vegetation and interception (and so evapotranspiration), which reduces moisture in the air (humidity). More precipitation also lands on hard, impermeable urban surfaces. So, surface runoff (overland flow) dominates and drains need to remove surface water quickly (Figure **1**).

Impacts on drainage basin storage areas

Drainage basins act as systems of inputs, transfers, outputs and stores. Urbanisation affects storage. For example:

- Management of urban river channels by dredging and **channelisation** will increase capacity.
- Reservoirs and ponds are permanent stores, but vulnerable to evaporation.
- Depression storage (e.g. puddles) is temporary.
- Interception and soil moisture storage is reduced owing to the replacement of vegetation by impermeable structures.

The urban water cycle: flood hydrographs

Urban hydrographs are 'flashy' ('peaky'), showing a rapid rise in discharge over a short period of time. This is because water mainly enters the river via surface runoff.

Urban flash flood risk is increased by:

- heavy rainfall/thunderstorms
- the high proportion of impermeable surfaces
- rapid runoff along roads, pavements and through drains
- the need to build on floodplains to meet the increased demand for housing
- culverts and urban rivers blocked with debris (particularly where bridges narrow the channel and so constrict the flow).

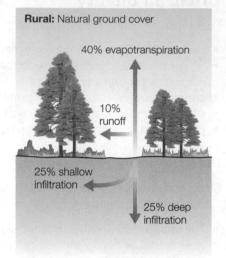

Rural: Natural ground cover
40% evapotranspiration
10% runoff
25% shallow infiltration
25% deep infiltration

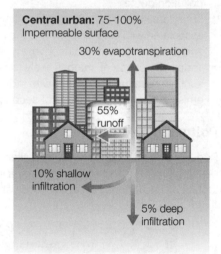

Central urban: 75–100% Impermeable surface
30% evapotranspiration
55% runoff
10% shallow infiltration
5% deep infiltration

⬣ **Figure 1** *Contrasting rural and urban catchment characteristics*

Figure 2 on page 155 of the student book is a hydrograph in an urban area.

Sixty second summary

- Urban areas have a higher rate of precipitation than rural areas, owing to warmer temperatures and higher levels of pollution.
- Urbanisation has significant impacts on drainage basin storage areas.
- There is an increased flood risk in urban areas because of the higher proportion of impermeable surfaces and the need to develop housing on floodplains to meet the growing demand for homes.

Over to you

Practise sketching and annotating a typical flood hydrograph for an urban area.

Student Book
pages 156–9

You need to know:

* about management of drainage within urban catchment areas
* Sustainable Drainage Systems (SuDS).

Management of drainage within urban catchment areas

Efficient, high-capacity urban drainage is essential if flooding is to be avoided – a management challenge that is met by hard- and soft-engineering approaches:

Hard engineering	Soft engineering
River straightening increases the gradient and speed of flow.	*Afforestation* increases interception and reduces throughflow and surface runoff. Evapotranspiration dissipates water that would otherwise end up in the river channel.
Natural levees can be heightened to increase capacity. *Embankments* are raised riverbanks using concrete or (sustainable) material dredged from the river bed.	*Riverbank conservation* by planting vegetation reduces lateral erosion, bank collapse and so silting (because roots bind and stabilise the banks).
Diversion spillways (flood relief channels) bypass the main channel – either during emergencies or permanently enhancing the environment by creating new wetlands.	*Floodplain zoning* restricts different land uses to certain locations on the floodplain (e.g. pasture nearest the channel).
Channelisation lines straightened river channels with concrete to reduce friction – improves flow rate and reduces silting.	*River restoration* returns the channel to its natural course – reversing past artificial management 'solutions' (see 3.19).

🔺 **Figure 1** *Hard- and soft-engineering management approaches*

Sustainable Drainage Systems (SuDS)

SuDS represent the ultimate in environmentally friendly, replication of natural drainage systems within a built environment. They hold back and slow surface runoff from any development and allow natural processes to break down pollutants.

🔺 **Figure 2** *Swales can be landscaped as attractive community, recreational spaces*

Techniques	Benefits
Swales – wide, shallow drainage channels; usually dry (Figure **2**)	Slows down surface water runoff
Permeable road and pavement surfaces (e.g. porous block paving)	Reduces the risk of sewer flooding during heavy rain – preventing water pollution
Infiltration trenches – gravel filled drains and filter strips	Recharges groundwater to help prevent drought
Bioretention basins – gravel/ sand filtration layers beneath reed beds to collect and filter dirty water	Wetland spaces provide valuable habitats for wildlife
Detention basins – excavated basins for water storage during flood events	Visual enhancement of landscape (amenity value) and groundwater recharge
Rain gardens – shallow depressions planted with flowers and shrubs	Creates green spaces for recreation (Figure **2**)
Green roofs – wildflower habitats with minimal runoff to gutters	Super-insulating properties

🔺 **Figure 3** *Techniques and benefits of SuDS*

Continued over ▶▶▶

A SuDS scheme: Lamb Drove, Cambourne, Cambridgeshire

This is an award-winning social housing development of 35 'affordable' homes in Cambridgeshire, a relatively low-lying county where flooding is a major concern. It:

- showcases practical and innovative sustainable water management techniques
- demonstrates that SuDs are a viable and attractive alternative to more traditional forms of drainage.

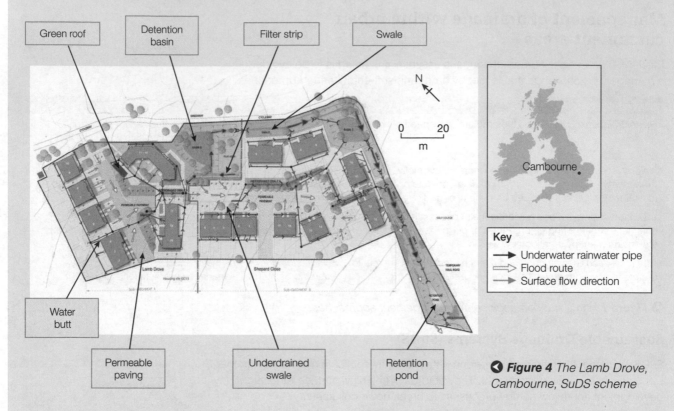

Green roof | Detention basin | Filter strip | Swale

Water butt | Permeable paving | Underdrained swale | Retention pond

Key
→ Underwater rainwater pipe
⇒ Flood route
→ Surface flow direction

 Figure 4 The Lamb Drove, Cambourne, SuDS scheme

SuDs measures adopted include:

- water butts to collect roof water for garden irrigation
- permeable paving allowing infiltration into porous storage zones
- swales to collect excess water, slow the flow and filter out pollutants
- detention basins and wetlands to temporarily store water during floods
- a retention pond for final water storage before being released to a drainage ditch beyond the development site.

Adoption of a *management train* at the site (a series of swales, detention basins and wetlands until water reaches the final retention pond) mimics natural drainage.

This controls the quantity of runoff incrementally by reducing flow rates and volumes, and improves water quality.

The closely monitored and appraised project is a proven success by:

- visually enhancing the development through attractive landscaping
- providing community spaces to further improve quality of life
- improving local biodiversity and ecology
- improving the quality of water leaving the site
- costing 10% less to construct and maintain than pipe drainage systems.

Sixty second summary

- Efficient urban drainage is essential to avoid flooding – but rivers must have the capacity to cope.
- Hard engineering can help to manage drainage within urban catchment areas, e.g. river straightening and diversion spillways.
- Soft engineering is also used to manage drainage, e.g. afforestation and floodplain zoning.
- The Lamb Drove project demonstrates the proven social, economic and environmental benefits of Sustainable Drainage Systems (SuDS).

Over to you

Outline the social, economic and environmental benefits of the Lamb Drove development.

You need to know:

- about restoration and conservation in damaged urban catchments
- about the aims and outcomes of a specific river restoration and conservation project

Student Book
pages 160–1

River Skerne, Darlington

River restoration is often impractical in urban areas because the built environment is so established. Gas and sewer pipes alongside the River Skerne had to be protected from erosion using soft revetments of willow mattress.

Most urban river restoration has to be partial, but such projects promote a greater emphasis on conservation within local land use and drainage management plans, providing a wide range of benefits:

- managing flood risk, by maintaining drainage
- conserving rich, native biodiversity
- stimulating investment to redevelop unsightly brownfield sites
- providing routeways and recreational opportunities for cyclists, anglers, bird watchers and ramblers
- promoting educational activities such as nature walks and wildlife gardening.

Big idea

River restoration is the removal of hard-engineering adaptations to restore meanders, wetlands and floodplains, returning flood management to nature.

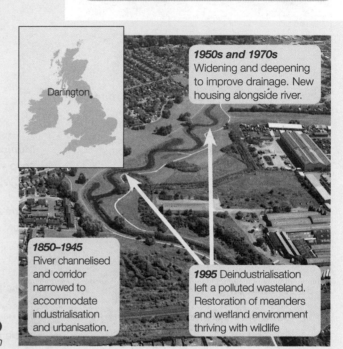

1950s and 1970s
Widening and deepening to improve drainage. New housing alongside river.

1850–1945
River channelised and corridor narrowed to accommodate industrialisation and urbanisation.

1995 Deindustrialisation left a polluted wasteland. Restoration of meanders and wetland environment thriving with wildlife

Darlington

Figure 1 ▶
Restoration of the River Skerne, Darlington

Sheffield city centre's Blue Loop

The Blue Loop Community Project follows the River Don and Sheffield and Tinsley Canal and has environmentally and socially transformed a derelict former industrial landscape into a multi-purpose conservation area.

The Sheffield Development Corporation, British Waterways, Natural England and the National Lottery all contributed expertise and funds to:

- rejuvenate a 13 km waterfront for cyclists and walkers
- restore natural ecosystems, and protect wildlife species and habitats
- clear non-native buddleia and Japanese knotweed
- engage local communities through volunteering, and promoting family events, festivals and recreation
- reduce flood risk by increasing floodplain wetland areas and adopting SuDS techniques.

 Sixty second summary

- Heavy industry in areas such as Darlington and Sheffield caused severe damage to the environment, including rivers.
- The River Restoration Centre restored a 2 km stretch of the River Skerne, attracting wildlife back to the area.
- Sheffield's Blue Loop Community Project illustrates the wide-ranging benefits of a multi-agency approach to urban river restoration and conservation.

 Over to you

Summarise the benefits of river restoration and conservation in urban areas (include at least one located example).

You need to know:
- about sources of urban waste
- how 'waste streams' relate to economic characteristics, lifestyles and attitudes
- the environmental impacts of alternative approaches to waste disposal.

Student Book
pages 162–7

Sources of urban waste

Urban waste is not simply domestic rubbish, but also comes from industrial and commercial activity.

 Big idea

Waste generation globally is growing exponentially. On average, people in HDEs produce 10 to 30 times more than those in LDEs.

All sources of urban waste

Industrial

Commercial (offices, retail stores, hotels, etc)

Institutional (schools, hospitals, etc)

Ashes (burning wood, coal, etc)

Bulky (domestic furniture and 'white goods', industrial pallets, etc)

Street sweeping (litter, dirt, leaves, etc)

Dead animals (that die naturally or are killed accidentally)

Animal and vegetable (food)

Construction and demolition

Municipal solid waste (MSW)

Domestic (residential e.g. empty containers, packaging, clothing, etc)

Municipal (market waste and abandoned vehicles, etc)

Type of waste

Hazardous (paint, spray cans)

Biodegradable (most can be recycled)

Recyclable materials (paper, glass, bottles, cans, certain plastics, etc)

Inert (construction and demolition debris)

Composite (drinks cartons, toys, etc)

Electrical and electronic ('white goods', TVs, computers, etc)

Toxic (pesticides, herbicides and fungicides)

Biomedical (expired prescription drugs)

🔺 **Figure 1** *Sources and types of urban waste*

Waste streams

The complete flow of waste from its domestic, commercial or industrial source, through to recovery, recycling or final disposal is known as a **waste stream**. In HDEs, this is increasingly regulated and managed.

However, in most LDEs and EMEs, indiscriminate and improper dumping of municipal solid waste (MSW) without treatment is common. This causes environmental problems because it:

- wastes recyclable resources (e.g. metals and glass)
- loses potential resources (e.g. energy from controlled incineration)
- contaminates land and water bodies with **leachates**
- pollutes air from burning (and methane from decomposition)
- endangers human health (e.g. respiratory problems, skin and other diseases).

Read about the waste-disposal problems in the Philippines on page 165 of the student book.

Global waste trade

The trade in waste – predominantly from HDEs to EMEs – enables its disposal, recycling or further treatment. But EMEs and LDEs do not always have safe recycling processes or facilities:

- workers process toxic waste with their bare hands
- untreated hazardous wastes are often dumped – with disastrous effects upon natural ecosystems.

Most waste electrical and electronic equipment (WEEE) is shipped to Asia and Africa to be processed and recycled (Figure **2**). Heavy metals and toxins leak from these discarded products into surrounding waterways and groundwater, poisoning the local people. Workers, scavengers in the dumps, and surrounding communities are all exposed to dangerous health risks.

Alternative approaches to waste disposal

Given the potentially harmful environmental impacts of urban MSW, effective management is essential.

- Sustainable MSW management can only be achieved with the *3Rs* (Figure **3**).
- Strict EU and UK government legislation regulates submarine dumping, landfill, incineration and recycling.
- There are strong arguments for and against landfill, incineration and recycling, but:
  - incineration and recycling tend to be the first thoughts in waste management
  - burial in landfill (albeit sealed and buried) remains its most usual fate either directly, or following extraction of recyclable waste in materials recovery facilities (MRFs).

Figure 2 *Workers in New Delhi, India dismantle obsolete computers and extract valuable materials such as nickel and copper with their bare hands*

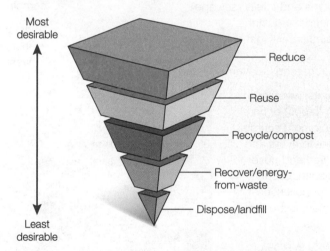

Most desirable — Reduce — Reuse — Recycle/compost — Recover/energy-from-waste — Dispose/landfill — Least desirable

Figure 3 *The* 3RS *– reduce, reuse and recycle*

 Read about the waste management in Bristol on page 167 of the student book.

 Sixty second summary

- Urban waste comes from a range of different sources, with people in HDEs producing 10 to 30 times more than those in LDEs.
- Municipal solid waste (MSW) is domestic and municipal refuse and can be categorized as hazardous, biodegradable or recyclable waste.
- The global waste trade is the international trade in the disposal, recycling or further treatment of waste.
- There are benefits and problems with waste management techniques such as landfill, incineration and recycling.
- Sustainable MSW management can only be achieved with the *3Rs* – reduce, reuse and recycle.

 Over to you

In a table, summarise **three** arguments both for and against **each** of landfill, incineration and recycling (Figure **7** on page 166 in the student book will help).

You need to know:

- about environmental problems in contrasting urban areas: atmospheric pollution, water pollution and dereliction
- strategies to manage these problems.

Student Book
pages 168–71

Atmospheric pollution

Look again at 3.15 and 3.16. Strategies to manage air pollution caused by domestic, industrial and vehicle emissions (e.g. clean air legislation, vehicle restrictions and technical innovations) are improving urban air quality, particularly in HDEs. But progress is not yet global, with a significant decline in air quality in larger cities in EMEs.

Water pollution

Urban water pollution comes from:

- domestic waste water from sinks, washing machines, bathrooms and toilets (sewage)
- effluent from industries
- leachates from illegal dumping and poorly managed landfill
- runoff from roads, pavements and roofs.

All pollutants, whether point (e.g. pipes) or non-point (e.g. runoff) sources, are a threat to rivers and groundwater, and pose a risk to health. Urban waste water collection and treatment is therefore essential. It is:

- obligatory in HDEs, with measurable improvements to river and ocean water quality
- less widespread in LDEs and EMEs, but some progress has been made as a result of investment by IGOs, such as the UN, and NGOs, such as Water Aid, and the Bill and Melinda Gates Foundation, to secure freshwater quality and supply.

Urban waste water treatment in the UK

There are three main types of waste water collection:

- *surface-water drainage* discharging directly to local waters
- *combined sewerage* that collects rainwater runoff and domestic, industrial, commercial waste water
- *foul drainage* that collects domestic waste water from premises.

Where there are no local environmental waters to which surface-water drainage can discharge, both surface-water and foul drainage may eventually connect to combined sewerage. But combined sewerage needs to:

- cater for storm water from peak seasonal wet weather
- include 'combined sewer overflows' to allow excess waste water to be discharged to local rivers to avoid sewers being overwhelmed.

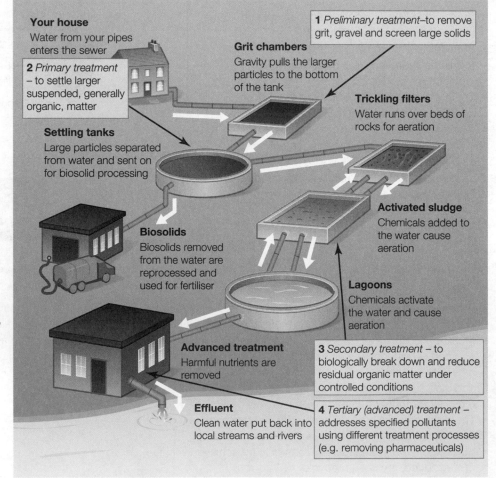

Your house
Water from your pipes enters the sewer

2 *Primary treatment* – to settle larger suspended, generally organic, matter

Settling tanks
Large particles separated from water and sent on for biosolid processing

Biosolids
Biosolids removed from the water are reprocessed and used for fertiliser

Grit chambers
Gravity pulls the larger particles to the bottom of the tank

1 *Preliminary treatment*–to remove grit, gravel and screen large solids

Trickling filters
Water runs over beds of rocks for aeration

Activated sludge
Chemicals added to the water cause aeration

Lagoons
Chemicals activate the water and cause aeration

Advanced treatment
Harmful nutrients are removed

3 *Secondary treatment* – to biologically break down and reduce residual organic matter under controlled conditions

Effluent
Clean water put back into local streams and rivers

4 *Tertiary (advanced) treatment* – addresses specified pollutants using different treatment processes (e.g. removing pharmaceuticals)

▲ *Figure 1* The four stages of sewage treatment

Dereliction

Derelict (neglected or abandoned) urban land results from:

- ageing and decay of buildings over time
- urban activities relocating to better and more profitable locations
- deindustrialisation.

Brownfield site development

The UK government focuses on a 'brownfield first' planning approach and so these unsightly sites represent a valuable urban resource. Their redevelopment for housing:

- helps address the pressing need for more homes
- improves the urban environment
- reduces urban sprawl and so protects green belts
- reduces commuting.

However:

- decontamination of contaminated ground and clearance of invasive species (e.g. Japanese knotweed) is both time-consuming and very expensive
- not all brownfield sites have appropriate physical access
- neighbouring land may still be used for incompatible purposes (e.g. a heavy industrial plant).

◈ Figure 2 *Urban dereliction can be hazardous and also be an eyesore*

Brownfield regeneration of Bristol's river docks area

Post-deindustrialisation, regeneration and redevelopment of the area has required cooperation between various **stakeholders** including the city council, private developers and the South West Regional Development Agency (Figure **3**).

Between 2006 and 2013, 94% of new housing in Bristol was built on brownfield sites and it is estimated that a further 30 000 new homes will be needed by 2026. The city council is confident that this can be achieved without using any greenfield sites.

Derelict land has been cleared and decontaminated for new housing, businesses, and cultural and leisure facilities

Decaying industrial buildings have been redeveloped for residential purposes

Several listed buildings have been preserved

 Figure 3 *Brownfield site regeneration of Bristol's river docks area*

Sixty second summary

- Pollution from energy use and vehicle emissions in urban areas lead to significant impact on human health, especially in EME cities.
- Urban water pollution comes from domestic and industrial sources; from point and non-point (diffuse) sources.
- Urban waste water collection and treatment is obligatory in HDEs, but less widespread in LDEs and EMEs. IGO and NGO investment is helping to secure (fresh) water quality and supply.
- Combined sewerage systems collect surface and foul water and waste water treatment involves four stages.
- Urban dereliction results from deindustrialisation. the ageing of buildings, relocation of urban activities elsewhere.
- Development in Bristol since 2006 has been mainly on brownfield sites.

Over to you

Copy **each** bullet point in the Sixty Second Summary and add a brief explanation and/or some evidence to support this point.

Student Book
pages 172–5

You need to know:

- about the environmental impact of major urban areas
- the dimensions of sustainability and the nature and features of sustainable cities
- about contemporary opportunities and challenges in developing more sustainable cities.

The impact of urban areas on local and global environments

Urban areas have a marked impact on both local and global environments. High densities of people and buildings:

- compete for space and consume vast quantities of water, energy and other resources
- create problems of air pollution, traffic congestion and waste disposal.

Furthermore, such are the needs and demands of urban populations in HDEs that their **ecological footprints** are more than ten-times those of people living in similarly sized urban areas in LDEs. So, any hopes of approaching global sustainability depend on developing more sustainable cities.

Figure 1 ◗

The ecological footprint is the area of land needed to provide the necessary resources and absorb the wastes generated by a community. The ecological footprint of the Tokyo metropolitan area has been calculated to be almost three times the land area of Japan as a whole!

Sustainable cities and liveability

Sustainability is all about making the best use of resources, protecting the environment, controlling growth and waste – all of which affect lifestyles in urban areas.

The concept of **liveability** describes the natural, physical, social and economic dimensions of sustainability in an urban context. It encapsulates the urban ideal of collectively improving everyone's quality of life both now and in the future. The truly liveable, sustainable city might well prove to be an impossible dream, but careful planning and management can certainly take us a long way towards it.

Carbon-neutral development: building structures that generate as much energy as they use, e.g. BedZED, Hackbridge, London

Greener built environments: using energy and water more efficiently, reducing MSW and managing it better (see 3.19 and 3.20), e.g. Quezon City, Philippines

Planned expansion: encouraging 'compact cities' by developing brownfield sites (see 3.21), e.g. Nantes, France

Nature and features of sustainable cities

Conserving buildings and open spaces: including urban river systems and ecosystems (see 3.19), e.g. Queen Elizabeth II Olympic Park, Stratford, London

'Liveable communities' with improved transport: developing infrastructure, networks and modes to meet demand (without increasing congestion and pollution); adopting affordable public transport, vehicle restrictions and technical innovations (see 3.16), e.g. Vitoria-Gasteiz, Spain

⬆ **Figure 2** *The nature and features of sustainable cities*

Contemporary opportunities and challenges in developing more sustainable cities

The goal of developing more sustainable cities inevitably faces challenges:

- *Political will*: long-term strategic planning is essential – involving all relevant **stakeholders**.
- *Globalisation*: the inter-connectedness of cities within the global economy has increased the power and influence of TNCs – so they must embrace the need for change.
- *Economic inequality*: sustainable cities must be inclusive – with economic incentives benefiting both the wealthiest and poorest.
- *Climate change*: must stimulate economic growth without increasing greenhouse gas emissions.

Freiburg, Germany: a sustainable city

Freiburg's goal of urban sustainability involves:

- Protected and enhanced green spaces. Black Forest woodlands cover 40% of the city. These woodlands are protected as nature conservation areas. Furthermore, the unmanaged River Dreisam (except for flood retention basins) provides natural habitats for flora and fauna.
- A sustainable water supply harnesses both rainwater harvesting and waste water recycling. SuDS include green roofs, permeable road and pavement surfaces, and widespread bioretention basins (see 3.18).
- MSW is reused and recycled – a biogas digester processes all food and garden waste, and energy for 28 000 homes is produced by incineration.
- Social dimensions of sustainability include the provision of 'affordable' energy-saving homes using renewable energy.
- Economic dimensions include the creation of 10 000 jobs in 1500 environmental businesses (e.g. in research, development and manufacture of solar panels).

 Figure 3 *Freiburg is a leading solar energy capital, with around 400 photovoltaic installations*

Sixty second summary

- Sustainability means meeting the needs of people today without compromising future generations.
- Modern developments often plan to be sustainable by being carbon neutral, conserving important buildings, using energy efficiently and incorporating public transport.
- There are a range of political, social, economic and environmental challenges to developing sustainable cities.
- Freiburg in Germany is a truly 'green city', incorporating green spaces, sustainable water supplies and reusing or recycling municipal solid waste.

Over to you

Rhyming definitions are easy to remember. 'Accessible places, natural spaces, minimal traces' was the winner of the Twitter competition to find an answer to '*What does liveability mean?*'.

Make your own rhymes for key themes on this page... and other pages.

You need to know:

- about an EME Olympic city – Rio de Janeiro, Brazil
- its patterns of economic and social well-being
- the nature and impact of physical environmental conditions.

Student Book
pages 176–9

Case Study

The geography of Rio de Janeiro – a city of contrasts

Rio is one of the most breathtaking cities in the world. As the host city of the 2014 FIFA World Cup and the 2016 Olympic and Paralympic Games, its global recognition was assured.

Figure 3 on page 177 of the student book details the four zones of Rio.

Rio is South America's top tourist destination, Brazil's former capital, and also a major port, financial and manufacturing centre. It is city of marked contrasts:

- hot, sunny (tropical) climate, rich culture and magnificent scenery including Sugar Loaf Mountain and Copacabana beach
- vast favelas, high unemployment and crime, traffic congestion, pollution, fatal landslides and extraordinary extremes of wealth.

It is divided into four main zones:

Centro – historic buildings; CBD; main shopping area
North – industrial and port area; airport; low-quality housing,
West – modern, luxury apartments; shopping attracts tourists; Olympic stadiums
South – main tourist area with beaches; wealthy but with largest favela in South America, Rocinha.

▲ *Figure 1* Location map of Rio de Janeiro

Rio's social and economic polarisation is best illustrated in the South Zone where *Rocinha*, South America's largest favela overlooks the main beaches, tourist hotels and high-rise luxury apartments (Figure **2**).

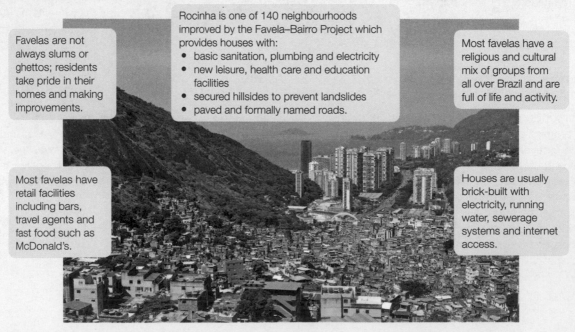

Favelas are not always slums or ghettos; residents take pride in their homes and making improvements.

Rocinha is one of 140 neighbourhoods improved by the Favela–Bairro Project which provides houses with:
- basic sanitation, plumbing and electricity
- new leisure, health care and education facilities
- secured hillsides to prevent landslides
- paved and formally named roads.

Most favelas have a religious and cultural mix of groups from all over Brazil and are full of life and activity.

Most favelas have retail facilities including bars, travel agents and fast food such as McDonald's.

Houses are usually brick-built with electricity, running water, sewerage systems and internet access.

▲ *Figure 2* Rocinha – Rio's largest and oldest favela with an estimated population of 150 000

Urban challenges and solutions in Rio

Housing, health and services

With over 100 000 in-migrants arriving each year, Rio faces acute housing problems:

- Many rent multiple-occupancy housing or occupy land as squatters, often hazardous environments (e.g. steep hillsides).
- One-third of 'informal' inhabitants have no access to sewers or electricity.
- Not everyone has piped water.
- Infant mortality is high and average life expectancy is in the 50s.

However, government improvement initiatives are proving to be successful (e.g. the Favela–Bairro Project).

▲ **Figure 3** *The cable car system services hillside favelas; it is the first true mass transit system in any slum in Rio*

Unemployment, underemployment and crime

- Unemployment is high (around 20%) and the poorly paid black market dominates.
- Poverty is rife, with the poorest 50% earning only 13% of Rio's income – virtually the same as the richest 1%!
- Corruption is widespread, with some favelas controlled by criminal gangs.
- Community policing (e.g. Pacifying Police Units – UPPs), strengthening of traditional industries (e.g. clothing) and encouragement of new economic activities (e.g. ICT and public transport) is helping.

Congestion, air and water pollution

- 40% of inhabitants live in Rio's suburbs; as a consequence four million cars jam the roads daily.
- Major road improvement projects (e.g. Rio–Niterói Bridge) have reduced congestion and air pollution significantly.
- Cable cars link favelas that were otherwise inaccessible (Figure **3**).
- The Metro system is being extended (Figure **4**).
- Tighter environmental controls are reducing raw sewage, industrial effluent, oil and landfill leachates in Guanabara Bay.

▲ **Figure 4** *The Rio de Janeiro Metro was opened and has been extended several times since then; the latest was into the West Zone for the 2016 Olympics*

 Sixty second summary

- Rio is South America's top tourist destination, and a major port, financial and manufacturing centre.
- There are several favelas in Rio, where thousands of people live in homes that often do not have electricity or basic sanitation.
- The rising population provides many challenges, including unemployment, crime, congestion and pollution.
- A number of solutions have been employed to reduce urban problems; new economic activity to reduce unemployment, community policing, tighter legislation, investment in infrastructure.

 Over to you

Summarise this topic in **five** sentences.

Case Study

You need to know:
- about an HDE Olympic city – London
- about the regeneration of London since the 1980s.

Student Book
pages 180–1

London – back from the brink

London is a world city and is the most visited city in the world. But it has not always thrived. By the early 1980s, huge swathes were in terminal decline owing to East End *counterurbanisation*, *deindustrialisation* and closure of the uncompetitive docks.

This prompted what was, at the time, the world's single largest economic and **property-led regeneration** – creating a *post-industrial economy* in the London Docklands via investment by London Docklands Development Corporation (LDDC) in:

- a secondary financial district (Figure **1**)
- a knowledge economy of high-value footloose business, e.g. ITC, media
- transport – improved infrastructure including the Docklands Light Railway, the Underground and City Airport
- brownfield development for 20 000 new homes, office and retail
- tourism.

But the regeneration was criticised for failing to provide affordable housing and employment. So, continued redevelopment, led by the London Thames Gateway Development Corporation (LTGDC), was **community-focused regeneration** characterised by social aims (e.g. 'affordable' homes) and sustainability (e.g. environmental enhancement of open spaces).

▲ *Figure 1* Canary Wharf financial district

The Olympic legacy

The redevelopment of the London Docklands was a factor in helping London win the competition to host the 2012 Olympic and Paralympic Games. Another was the promise that it would be the 'greenest' and 'most sustainable' Games in history. The aim was to achieve **sports-led regeneration** creating a lasting legacy of jobs, homes and environmental improvements.

Housing – the athletes' village is now 2800 homes, 1200 of which are 'affordable'.

Transport – infrastructure includes roads and Stratford International (Eurostar) rail station.

The Olympic legacy – sports-led regeneration

Employment – the International Broadcast Centre is now iCITY supporting 4600 sustainable, high-tech jobs with a further 2000 in the supply chain.

Environment – River Lea and adjoining canals cleaned up, towpaths opened up to walkers/cyclists and 4000 trees planted, creating new habitats e.g. Queen Elizabeth II Olympic Park.

▲ *Figure 2* The Olympic legacy

 Sixty second summary

- London is a thriving world city and the most visited in the world.
- Deindustrialisation and counterurbanisation led to dramatic decline by the 1980s, creating urban challenges such as *deprivation* and unemployment, particularly in the East End.
- The property-led regeneration of the Docklands is one of several initiatives to boost employment, transport, housing and the environment in East London.
- The 2012 Olympic and Paralympic Games were planned to achieve sports-led regeneration in the six London boroughs that hosted the Games.

 Over to you

Make sure you are familiar with and understand the key terms of this topic (identified in **bold** and *italics*).

4 Population and the environment

Your exam

(AL) *Population and the environment* is an **optional topic**. You must answer **one** question in Section C of Paper 2: Human geography, from a choice of **three**: *Contemporary urban environments* or *Population and the environment* or *Resource security*.

Paper 2 carries 120 marks and makes up 40% of your A Level. Section C carries 48 marks.

Specification subject content (specification reference in brackets)

Either tick these boxes as a record of your revision, or use them to identify your strengths and weaknesses

Section in Student Book and Revision Guide	1 ☹	2 😐	3 🙂	Key terms you need to understand Complete the key terms (not just the words in bold) as your revision progresses. 4.1 has been started for you.
Introduction *(3.2.4.1)*				
4.1 Population and the environment themes				*Exponential growth, carrying capacity, population parameters, physical factors, development processes*
Environment and population *(3.2.4.2)*				
4.2 Patterns of food production and consumption				
4.3 Agricultural systems and productivity				
4.4 Climate and climate change				
4.5 Soils and human activities				
4.6 Soil problems and management				
4.7 Food security				

Your revision checklist

Environment, health and well-being *(3.2.4.3)*				
4.8 Global health				
4.9 Health and morbidity in the UK				
4.10 What influences health and well-being?				
4.11 The relationship between place and well-being				
4.12 Disease and the physical environment				
4.13 Malaria: the geography of a biologically transmitted disease				
4.14 Malaria: Millennium Development Goals and eradication				
4.15 Asthma: the global impact of a non-communicable disease				
4.16 Asthma: management and mitigation to maximise health and well-being				
Population change *(3.2.4.4)*				
4.17 Natural population change				
4.18 Models of natural population change				
4.19 Population structure				

4.20 Factors of natural population change			
4.21 Migration change			
4.22 International migration			
4.23 Implications of migration to Australia			

Principles of population ecology and their application to human populations *(3.2.4.5)*

4.24 Environmental constraints on population growth			
4.25 Balancing population and resources			
4.26 How will global population change?			

Global population futures *(3.2.4.6)*

4.27 Health and environmental change			
4.28 Global population – the future			

Case studies *(3.2.4.7)*

4.29 Population change in Iran			
4.30 Relationship between place and health			

You need to know:

- about physical factors affecting population, key population parameters, the role of development processes, and global patterns of population growth.

*Student Book
pages 186–7*

A growing population

- Population has experienced **exponential growth** since the 1800s (Figure **1**).
- The world population was estimated to be 7.6 billion people in June 2018.
- Population is growing in most countries of the world (Figure **2**).
- There is no single factor that influences population growth – soil, climate, resource availability, development processes and human behaviour may all influence the growth rate of a population.
- Eventually, we may exceed the **carrying capacity** of the planet.

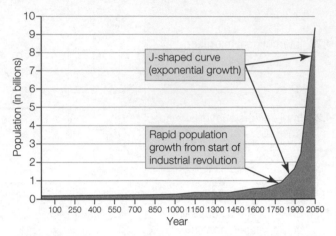

 Figure 1 *Growth of the world's population*

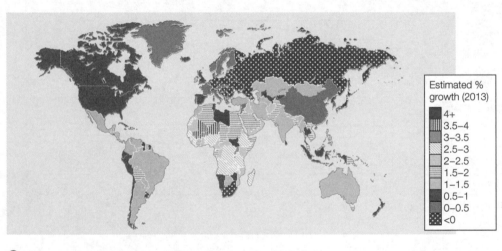

 Figure 2 *Estimates of population growth rates (2013)*

- Global average population is growing at around 1% annually
- Rapid population growth in Africa
- Population is decreasing in some areas – mainly in Eastern Europe (purple areas)

Niger – an example of how physical factors impact on population growth

According to the UN's Human Development Index (2015), Niger is the world's poorest country.

- Environmental challenges are a cause of poverty (unreliable rainfall, desertification, rural to urban migration).
- Niger has the world's highest fertility rate – children support their families by working the land and, when older, finding jobs in towns to the south.
- Overgrazing and drought has caused grazing pasture to disappear.
- Niger relies on foreign aid and food imports – it is *food-insecure*.

Sixty second summary

- Global population is continuing to increase, particularly in the poorest countries of Africa.
- Environmental factors that influence population change include climate, soils, resource distribution and water supply.
- Development processes also influence population change, particularly where a larger family is considered an economic necessity.
- Niger suffers significant environmental challenges that act as barriers to development.

Over to you

Using Figures **1** and **2**, practise describing and explaining the pattern of population change in your projected lifetime (to the year 2100).

You need to know:
- about global and regional patterns of food production and consumption.

Student Book
pages 188–9

Patterns of food production and consumption

Global and regional patterns of food production and consumption demonstrate remarkable contrasts:

- Most HDEs have enough farmland to provide the food they need, with significant surpluses to export.
- Half of all LDEs lack sufficient farmland and technology to be self-sufficient and are too poor to import.
- Enough food is currently produced to feed the world. The 2015 **Millennium Development Goal** (MDG) to halve the proportion of hungry people was almost met.
- One-third of all food produced worldwide is wasted.
- One in nine people worldwide suffer from **chronic undernourishment** – a diet lacking in protein, energy, vitamins and minerals (Figure **1**).

Malnutrition also includes *overnutrition* and, globally, eating too much is now a more serious health risk than eating too little (Figure **2**).

An oversize epidemic?

Although preventable, worldwide **obesity** has doubled since 1980. It is associated with:

- 'diseases of affluence' in HDEs linked to over-indulgence and the pressured way of life
- low-income earners, with imbalanced carbohydrate dominant and/or processed food diets
- diseases such as coronary heart disease, type 2 diabetes and cancers.

Solutions involve making people's choices of healthier foods and physical activity easier, and both governmental and societal pressure on the food industry to:

- reduce non-saturated fats, sugar and salt in processed foods
- ensure that healthy food such as fruit and vegetables are available and affordable
- restrict marketing of foods high in fats, sugars and salt
- encourage physical activity in the workplace.

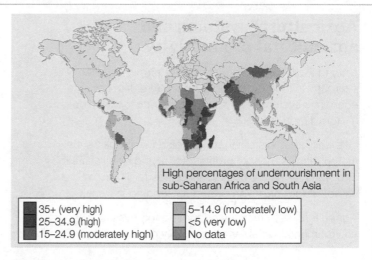

High percentages of undernourishment in sub-Saharan Africa and South Asia

- 35+ (very high)
- 25–34.9 (high)
- 15–24.9 (moderately high)
- 5–14.9 (moderately low)
- <5 (very low)
- No data

Figure 1 *Percentage of undernourishment in the population (2014–16)*

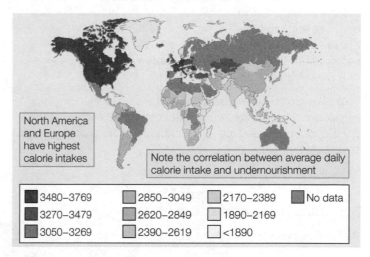

North America and Europe have highest calorie intakes

Note the correlation between average daily calorie intake and undernourishment

- 3480–3769
- 3270–3479
- 3050–3269
- 2850–3049
- 2620–2849
- 2390–2619
- 2170–2389
- 1890–2169
- <1890
- No data

Figure 2 *Average daily calorie intake per capita (2011)*

Sixty second summary

- Global and regional patterns of food production and consumption demonstrate remarkable contrasts.
- Undernutrition will persist until food is more evenly distributed and waste reduced.
- Globally, overeating is now a more serious health risk than not eating enough.
- Obesity is preventable by making the choice of healthier foods and regular physical activity easier.

Over to you

Write down **three** ways that patterns of food consumption globally could be improved.

You need to know:

- about agricultural systems and productivity
- their relationship with climate, relief and soils.

Student Book
pages 190–3

Agricultural systems and agricultural productivity

Around 2 billion people globally are employed directly or indirectly in food production. Agriculture, like any other economic system, involves inputs, processes and outputs (Figure **1**).

Like any other system there are also feedbacks, such as the reinvestment of profits into buying additional land and equipment, the use of manure as fertiliser and the use of fodder crops produced for feeding livestock.

Agricultural systems vary according to physical environmental variables (e.g. mainly climate and soils but also relief).

Human factors are also important. Socio-economic, behavioural, cultural, scientific and even political influences have a significant bearing on agricultural land use and productivity. It is important, therefore, to appreciate that farming practices will demonstrate great variability and frequently contain elements of more than one type (Figure **3**).

There will also be changes to the system – either physical environmental (e.g. impacts of droughts and floods), or human (e.g. changes in demand or political policies).

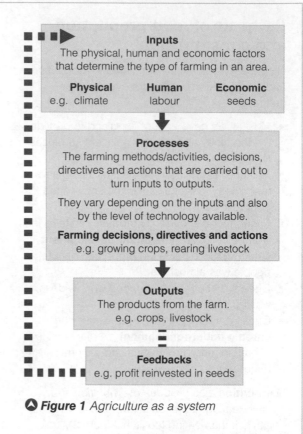

Inputs
The physical, human and economic factors that determine the type of farming in an area.

Physical	Human	Economic
e.g. climate	labour	seeds

Processes
The farming methods/activities, decisions, directives and actions that are carried out to turn inputs to outputs.

They vary depending on the inputs and also by the level of technology available.

Farming decisions, directives and actions
e.g. growing crops, rearing livestock

Outputs
The products from the farm.
e.g. crops, livestock

Feedbacks
e.g. profit reinvested in seeds

⊗ *Figure 1 Agriculture as a system*

Physical environmental factors influencing agriculture

Climate, soils and relief are the most important physical environmental factors affecting agricultural productivity.

- *Temperature* dictates the growing season as determined by the number of frost-free days.
- *Precipitation* is a factor in determining the water supply. Its effectiveness depends on temperatures and therefore rates of evapotranspiration. Seasonal distribution of rainfall is particularly important.
- *Wind* and storm frequency restricts cultivation of grain crops.
- *Soil quality* is crucial and determined by factors such as depth, texture, structure, mineral content, pH, aeration, capacity to retain water and vulnerability to **leaching**.
- *Relief variables* interrelate with the climate and soil factors above. They include altitude, angle of slope and aspect (Figure **2**)

◀ *Figure 2*
Warm south-facing aspect maximises insolation for these vineyards in the Rhône Valley, Switzerland – in contrast to the woodland covering the north-facing slopes

Agricultural system	Examples
Arable farming is the farming of cereal and root crops, usually on flatter land where soils are of a higher quality. Arable farming can be subsistence or commercial.	*Slash-and-burn* – shifting cultivation in Latin America, Africa and south-east Asia. Once a plot loses fertility people clear another, only returning when the original vegetation has regenerated naturally. *Commercial* – potato cultivation in Hertfordshire, UK.
Commercial farming typically involves farmers and **agribusinesses** maximising profits by specialising in single crops (monoculture) or raising one type of animal. Commercial farming will often involve high investment of capital into land, contractors, machinery, agrochemicals and animal welfare.	*Arable* – grain cultivation in North America, such as the Canadian winter wheat harvest, and tea plantations in East Africa. *Pastoral* – cattle ranching on the South American Pampas. Today only 20% of Argentinean beef is reared sustainably on grass.
Extensive farming uses low inputs of labour, machinery and capital, but usually involves large areas of land; yields per hectare are consequently low. It is therefore the opposite of intensive farming.	Hill sheep farming in UK upland areas, such as Snowdonia, the Lake District and Yorkshire Dales – much of the character of these areas is maintained through sheep farming, though the farms often run at the margins of profitability.
Intensive farming involves high investment in labour and/or capital such as machinery, glasshouses and irrigation systems – even hydroponics. It produces high yields per hectare from often small areas of land.	Fruit, flower and vegetable production (horticulture) in south-west England and the Netherlands.
Mixed farming is, as the name implies, the production of both arable crops and livestock. Commercially sensible in allowing flexibility (including diversification into farm shops and leisure activities such as 'glamping') it is particularly suited for spreading risk.	Mixed farming is the most common form of agriculture in the UK.
Pastoral farming involves livestock rearing and can be subsistence or commercial.	*Subsistence* – nomadic pastoralism (herding of cattle, goats and camels) in semi-desert regions of West Africa. *Commercial* – sustainable beef cattle ranching on the South American Pampas.
Subsistence farming involves the direct production of sufficient food to feed the family or community involved, with any excess produce sold or bartered.	Nomadic pastoralism in West Africa, slash-and-burn shifting cultivation in Latin America, Africa and south-east Asia, and intensive rice cultivation in Asia are examples of the few purely subsistence farming systems left. But 'mainly' subsistence is still important.

⊙ **Figure 3** *Examples of agricultural systems*

Sixty second summary

- Around 2 billion people globally are employed directly or indirectly in food production.
- Agricultural systems involve inputs, processes, outputs and feedbacks.
- Of all the physical environmental factors affecting agricultural productivity, climate and soils are the most important.
- Arable, commercial, extensive, intensive, mixed, pastoral and subsistence agricultural systems can be linked – for example, commercial arable farming takes place in Manitoba, Canada.

Over to you

Using Figure **1** as a template, draw systems diagrams for **three** contrasting agricultural systems. For example:

- an intensive arable system in an HDE
- an extensive subsistence system in an LDE
- and an agribusiness in **either** an HDE **or** LDE.

Student Book
pages 194–5

You need to know:

- about the characteristics and distribution of two major climatic types and how they influence human activities and numbers
- how climate change affects agriculture.

Two contrasting climates: tropical monsoon and polar tundra

The tropical monsoon climate and polar tundra climate contrast in many ways. Figure **1** outlines some of them.

	Tropical monsoon climate	Polar tundra climate
Distribution	• Typical of south and south-east Asia	• Tundra regions of northern hemisphere, including areas of Canada and Greenland
Key characteristics	• Wet summer season – high temperatures and huge amounts of rain • Dry winter season – cooler temperatures and drought	• Windy with limited precipitation and very cold temperatures • Climate restricts agriculture
Geographical significance	• Affects over half the world's population in over 21 Asian countries	• Polar climates cover more than 20% of the Earth's surface
Human activities	• Labour-intensive subsistence rice cultivation • 'Wet' rice on flat, alluvial flood plains • 'Dry' rice on irrigated hillside terraces	• Sustainable support of low numbers of nomadic **indigenous** Inuit (Canada, Greenland) and Sami (northern Europe) – hunting and fishing at subsistence levels

🔼 *Figure 1 Comparison of the features of the tropical monsoon and polar tundra climates*

The effects of climate change

Tropical monsoon climate

- Climate change will result in less predictable weather conditions.
- Variations in precipitation jeopardise traditional rice production.
- Research into less water-intensive methods of cultivation (e.g. *ClimaRice* in southern India) has demonstrated increased yields but greater need for herbicides.

Polar tundra climate

- For more than 30 years, northern high latitudes have experienced significant warming, which is predicted to continue leading to a warmer, wetter climate.
- Widespread thawing of permafrost, glacial retreat, a shorter snow season and reduced sea ice will increase rates of coastal erosion and slope instabilities.
- The sustainability of existing settlements, infrastructure and lifestyles are threatened, but transport, tourism and mineral exploitation may benefit. Traditional activities will inevitably change.

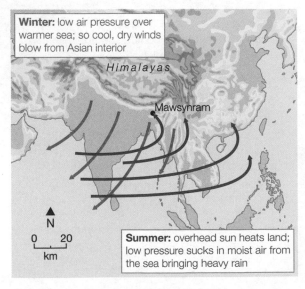

Winter: low air pressure over warmer sea; so cool, dry winds blow from Asian interior

Himalayas

Mawsynram

N

0   20
km

Summer: overhead sun heats land; low pressure sucks in moist air from the sea bringing heavy rain

🔼 *Figure 2 The tropical monsoon climate in south-east Asia*

 Sixty second summary

- Intensive subsistence rice cultivation depends on the seasonal rains of the tropical monsoon climate.
- Tropical monsoon climate change is prompting research into less water-intensive methods of rice cultivation.
- The harsh polar tundra climate and associated vegetation have sustainably supported indigenous people at subsistence levels for thousands of years.
- Polar climate change is already significant, with warming causing landscape instabilities and threatening lifestyles.

 Over to you

Test your memory by looking at this page again in two days' time.

Student Book
pages 196–7

You need to know:

- about the characteristics and distribution of two key zonal soils
- their relationship with human activities.

Tropical red latosol

Tropical red latosol is the deep zonal soil of the tropical equatorial rainforest biome, yet it is inherently infertile.

- The hot, wet, humid climate promotes perfect year-round growing conditions for leaf-shedding deciduous trees.
- The constant supply of leaf litter decomposes rapidly into humus, supplying nutrients to support sustainable new growth.
- But forest clearance stops nutrient cycling and the soil quickly becomes exposed to leaching and erosion by gulleying during the heavy, daily rainstorms.

For generations humans have lived sustainably within tropical equatorial rainforests. But recent decades have seen deforestation:

- to create land for settlement and associated infrastructure
- to create land for ranching, cash-cropping and plantations
- for hardwood timbers
- to provide access for mineral exploitation.

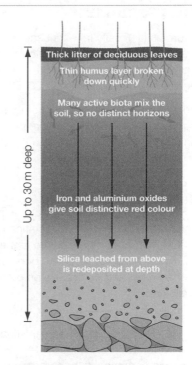

Figure 1 *Profile of the deep, iron-rich tropical red latosol*

Podsol

Podsol is the shallow zonal soil of the taiga biome, the vast belt of subarctic climate across Northern America and Eurasia.

Prolonged harsh winters and cool summers restrict the vegetation to boreal coniferous forest, which creates largely infertile soils.

- Low temperatures and shading coniferous trees ensure that precipitation provides a surplus over evapotranspiration – allowing downward infiltration, percolation and leaching.
- Coniferous evergreens return few nutrients when the leaves (needles) eventually fall.
- The cycling of nutrients is very poor.

Most podsols lie beneath the North American and Eurasian taiga, which is slowly disappearing.

Logging in North America is now managed but, since the fall of the Soviet Union, commercially lucrative logging of areas of Eurasia has been encouraged.

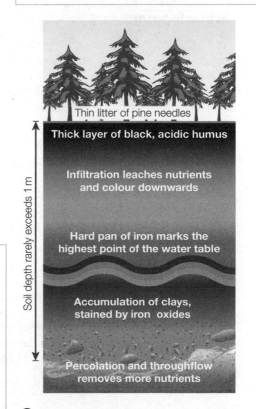

Figure 2 *Profile of distinctive horizons in a typically shallow podsol*

Sixty second summary

- The well-mixed tropical red latosol is the zonal soil of the tropical equatorial rainforest biome.
- Tropical red latosols quickly degrade following deforestation.
- The distinctively layered podsol is the zonal soil of the gradually disappearing taiga biome.

Over to you

Create a Venn diagram to compare and contrast key characteristics of tropical red latosols and podsols.

You need to know:
- about what problems affect soil
- how soil management relates to agriculture.

Student Book
pages 198–201

Soil problems

Mature soils usually take thousands of years to form, but are being lost at between 10 and 40 times the rate at which they can be naturally replenished. Nearly half of all soils are being degraded and without management, food production could decline by 30% over the next 30–50 years.

Perhaps surprisingly, *waterlogging*, *salinisation* and *structural deterioration* are more of a problem than rain splash, sheet wash, rill and gulley erosion (Figure **1**).

Soil erosion

Soil erosion by wind and water is often blamed on the removal of natural vegetation cover – leaving the ground directly exposed to the elements. Erosion is inevitable where poor land management, often provoked by population pressure, has allowed:

- deforestation
- overgrazing and overcultivation
- reduced fallow periods
- use of monoculture across large fields
- deep ploughing up and down slopes.

⚠ **Figure 1** *Severe gulley erosion following excessive winter rainfall in Devon, February 2014*

Waterlogging

Waterlogging (soil saturation) provides insufficient oxygen for plant roots to respire, causing them to die. It also damages soil structure and, in hot climates, provides excellent breeding grounds for mosquitoes (Figure **2**). Waterlogging occurs where:

- rainfall exceeds the rate that soils can absorb or the atmosphere can evaporate it
- gentle relief and basins restrict throughflow and encourage accumulation of water
- soils include an impermeable clay layer or iron pan
- excessive irrigation water is used to flood fields.

⚠ **Figure 2** *Waterlogged field in Thailand after heavy monsoon rain in 2011*

Salinisation

Salinisation (weathering of soil minerals and releasing salts) need not be a problem, providing soils are well drained. But waterlogging brings dissolved salts towards the surface, where evaporation leaves a crust of concentrated salt, which only salt-tolerant crops like cotton can withstand.

Structural deterioration

Structural deterioration by salinisation, or heavy machinery, can seriously degrade a soil. For example, a clay soil with a crumbly structure which is easy for air, water, and roots to move through, may be made impenetrable by compaction.

Soil management

Soil management is all about conservation – in its simplest form it is preserving and protecting vegetation cover. For example, afforestation and reforestation provide the best long-term solution to soil erosion because once the trees have grown, their foliage shades the soil from the Sun and intercept rainfall. Their roots also help to bind the soil together and reduce surface runoff.

But agriculture is the purposeful tending of crops and animals (not trees) and soil conservation also includes protecting soils from waterlogging, salinisation, reduced fertility and, at worst, exhaustion.

Therefore, soil management must include water management as well as the preservation of vegetation cover – both are integral to good agricultural land management.

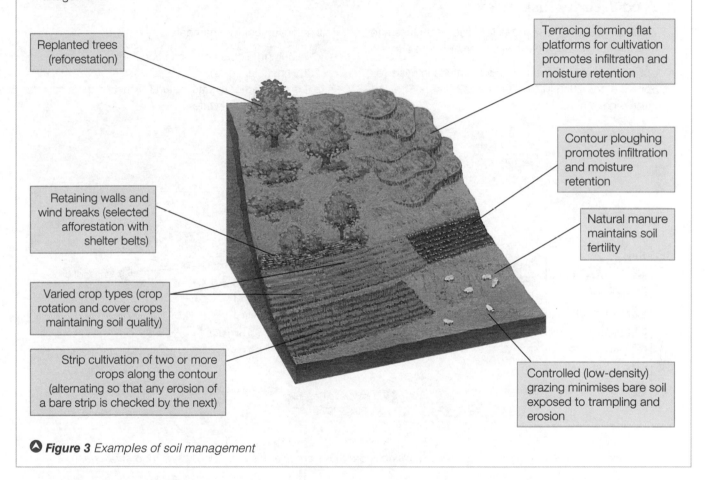

Replanted trees (reforestation)

Terracing forming flat platforms for cultivation promotes infiltration and moisture retention

Contour ploughing promotes infiltration and moisture retention

Retaining walls and wind breaks (selected afforestation with shelter belts)

Natural manure maintains soil fertility

Varied crop types (crop rotation and cover crops maintaining soil quality)

Strip cultivation of two or more crops along the contour (alternating so that any erosion of a bare strip is checked by the next)

Controlled (low-density) grazing minimises bare soil exposed to trampling and erosion

 Figure 3 *Examples of soil management*

 Sixty second summary

- Nearly half of all soils are being degraded and, without management, food production will decline.
- Soil erosion is inevitable where poor land management has allowed overgrazing, overcultivation, reduced fallow periods and deforestation – leaving it vulnerable to wind and water.
- Soil degradation by waterlogging, salinisation and structural deterioration are more of a problem than rain splash, sheet wash, rill and gulley erosion.
- Soil management is concerned with the preservation of vegetation cover, but must also include water management.

Over to you

Create a summary table directly relating soil problems to their management.

You need to know:

- about strategies to ensure food security.

Student Book
pages 202–5

What is food security?

This refers to people always having enough safe, nutritious food to maintain a healthy and active life. It must be *available* (consistently, in sufficient quantities), *accessible* (regularly, e.g. via purchase, home production, even food aid) and *utilised* positively (stored and cooked hygienically).

Food Security Risk Index

The Food Security Risk Index identifies countries and regions whose food security is at risk (Figure **1**).

But although hunger remains in countries in crisis (e.g. Somalia), the 2015 Millennium Development Goal (MDG) targets have almost been met.

Full achievement was hampered by:

- challenging global economic conditions (e.g. the 2008 financial crisis)
- extreme weather events, natural disasters, political instability and civil strife.

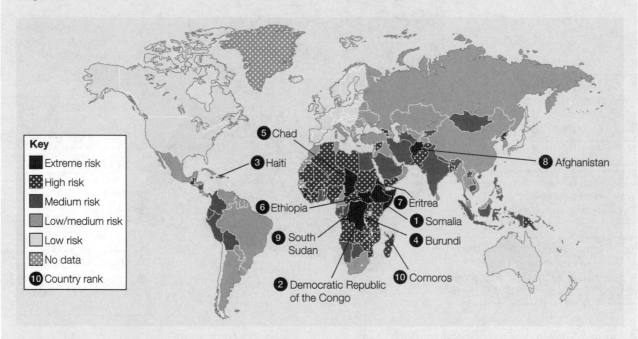

Key
- ■ Extreme risk
- ▨ High risk
- ■ Medium risk
- ■ Low/medium risk
- □ Low risk
- ▨ No data
- ❿ Country rank

⑤ Chad
③ Haiti
⑥ Ethiopia
⑨ South Sudan
② Democratic Republic of the Congo
⑦ Eritrea
① Somalia
④ Burundi
❿ Comoros
⑧ Afghanistan

⬆ **Figure 1** *Food Security Risk Index (2013), which infers those areas that suffer hunger and food shortages*

Strategies to ensure food security

The near-achievement of the MDG hunger targets shows that elimination of hunger in your lifetime is not unrealistic. Important lessons for the future have been learnt from the MDGs. For example:

- *Improved agricultural productivity*, especially by small and family farmers, leads to reduced hunger and poverty. This has been achieved through:
  - the Green Revolution
  - irrigation in drylands, swamp drainage and land clearance in LDEs and EMEs
  - grants, subsidies and guaranteed prices in HDEs (e.g. the EU's CAP).

The emphasis is now more on the balance between food production and environmental stewardship. It is also about a rising prevalence of obesity.

- *Economic growth* is always beneficial because it expands the fiscal revenue base necessary to fund improvements.
- *Expansion of social protection.* The FAO estimates these measures (e.g. food vouchers, health insurance and school meal programmes) help 150 million worldwide, but there is still much to do. More than two-thirds of the world's poor still do not have access to regular and predictable forms of social support.

The Green Revolution

The Green Revolution refers to a range of innovations that began in the 1960s.

- The development of hybrid, high-yielding varieties (HYVs) of crops (e.g. rice, Figure **2**).
- Cross-breeding of animals to improve their tolerance to difficult environmental conditions (e.g. aridity).
- Increased use of agrochemicals (e.g. pesticides, Figure **3**).
- Developing synthetic hormones to control plant sizes and growth rates.
- Increasing water control and irrigation schemes.
- Improving crop storage, handling, and processing.
- Mechanisation and land reform, including reorganisation of fragmented plots.
- Soil conservation (e.g. contour strip cropping).

But the Green Revolution introduced a number of economic and social problems:

- Only richer farmers could afford mechanisation which, in turn, increased unemployment and rural depopulation.
- Poorer farmers took out loans to buy HYVs and agrochemicals, but debt defaulting forced land sales.
- Use of agrochemicals could be dangerous, although the production of them has created jobs.
- Education was needed to ensure productive cultivation of HYVs, given their special agrochemical and irrigation requirements.

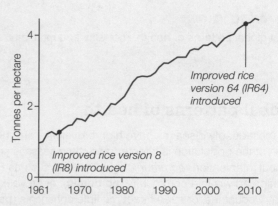

 Figure 2 *Global rice yields showing impact of HYV 'miracle' rice varieties*

 Figure 3 *Crop spraying in Punjab, India, has helped double farm yields*

The Gene Revolution

Adoption of *genetic modification (GM)* has been rapid, though controversial – GM crops are still banned in nearly 40 countries. It involves transferring genetic DNA between plant species to improve resistance to drought, specified pests or diseases. But concerns about the, largely unknown, implications for human health and the environment are widespread.

Seed collection from wild species that closely relate to common food crops is being undertaken by scientists (*Crop wild relatives research*). This approach seeks to conserve natural genetic diversity and could help to make future crops more resilient in the face of climate change and other threats.

 Sixty second summary

- Food security analysts consider food availability, food access and food utilisation.
- Strategies to ensure food security include improved agricultural productivity, economic growth and social protection measures.
- The Green Revolution has ensured massive increases in yields, but caused economic and social problems.
- The development of GM crops has proved controversial.
- Scientists are studying crop wild relatives (and creating a seed bank) to secure crop resilience naturally.

Over to you

Practise outlining:

(a) the meaning of food security,
(b) where food security is of greatest risk, and
(c) **advantages** and **disadvantages** of **both** the Green and Gene Revolutions.

You need to know:

- about global patterns of health, mortality and morbidity.

Student Book
pages 206–7

Global patterns of health

The most deadly diseases have high prevalence and high incidence rates – much of the population is infected and the disease is spreading quickly. Without intervention, *communicable* diseases such as Ebola may become widespread within a community (**epidemic**); or prevalent over a whole country/countries, or even affect multiple continents (**pandemic**).

The Ebola epidemic in West Africa

Ebola was first recognised in tropical sub-Saharan Africa in 1976. The 2013–16 epidemic killed over 10 000 people.

- The **index case** was in Guinea, in 2013.
- The disease quickly spread to Liberia and Sierra Leone (Figure **1**).
- European countries treated patients who had contracted the virus from the **endemic** West Africa.
- By 2016 the WHO declared Liberia to be 'Ebola-free'.

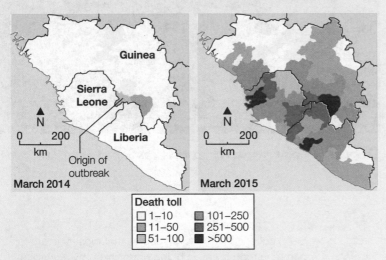

March 2014　March 2015

Death toll
- □ 1–10
- ▨ 11–50
- ▨ 51–100
- ▨ 101–250
- ▨ 251–500
- ■ >500

🔺 **Figure 1** *The spatial diffusion (spread over time and space) of the Ebola virus was rapid*

Responses to the epidemic

- The WHO was criticised for being slow to react, while volunteers working for NGOs risked their lives to treat victims.
- In the UK, emergency meetings of **COBRA** drew up contingency plans to protect the public (Figure **2**).
- In West Africa, policies of quarantine and isolation encouraged schools to close for six months.

Key terms

Health: Physical, mental and social well-being, not just the absence of disease or infirmity. Life expectancy is one useful indicator. **Disability-adjusted life years (DALYs)** associated with a population or disease is another (see 4.16).

Mortality: Being subject to death, mortal. A common indicator is *death rate* (see 4.17).

Morbidity: Illness or poor health. Indicators include:

- *Prevalence rate* – total number of cases of a disease (at a given time) divided by total population.
- *Incidence rate* – rate at which persons become ill. A measure of the number of new cases of an illness.

Ebola in West Africa
Information for the public

If you have returned from **Guinea, Liberia** or **Sierra Leone** or **cared for someone with Ebola** in the past **21 days**

and

You have a **fever** or **feel unwell**

Without touching anyone, **tell a member of staff** or **call 111**

Ebola facts:
- the risk of Ebola to the UK public is very low
- people with early symptoms (such as fever or sore throat) are **unlikely** to spread Ebola
- Ebola is not spread through the air
- however, someone with Ebola can be **infectious** if they are suffering from diarrhoea, vomiting or bleeding

For more information visit www.gov.uk/phe or www.nhs.uk/ebola

🔺 **Figure 2** *Posters highlighted the risks of travel to West Africa at this time*

 Sixty second summary

- Health means more than absence of disease or infirmity; it is about physical, mental and social well-being.
- The deadliest diseases carry the risk of epidemic, even a pandemic at a global scale.
- In the first twelve months of the 2013–16 epidemic, Ebola killed more than 10 000 people.

Over to you

Write your own *taboo cards* based on terms from the global health topic. For each term, note down **three** or **four** associated words, that may not be said in the game.

Student Book
pages 208–9

You need to know:

- about regional variations in health and morbidity
- factors that influence these variations.

Variations in health and morbidity in the UK

Despite it being an HDE, there are significant variations in average life expectancy within the UK.

- There is a health 'north–south divide'. With increasing distance from London and the south-east of England, there is decreasing life expectancy.
- Variation between the four nations that make up the UK. The average life expectancy of men in Scotland is consistently lower than those in England (Figure **1**).
- There is even variation within London. Life expectancy decreases by 12 years between communities just 20 minutes apart on the London Underground's Central line.

	1992–4	2002–4	2012–14
Males			
United Kingdom	**73.7**	**76.2**	**79.1**
England	73.9	76.4	79.4
Wales	73.4	75.8	78.4
Scotland	71.7	73.8	77.1
Northern Ireland	73.0	75.8	78.3
Females			
United Kingdom	**79.0**	**80.7**	**82.8**
England	79.2	80.9	83.1
Wales	78.9	80.3	82.3
Scotland	77.3	79.1	81.6
Northern Ireland	78.7	80.6	82.3

⬤ **Figure 1** *England tends to have a higher average life expectancy that the other three UK nations*

Factors that influence health and morbidity

Occupation

- Those living in poverty (e.g. in **deindustrialised** regions) are likely to suffer poorer health.
- Type of occupation can influence health, e.g. working at home as a young carer may affect mental health.

Lifestyle

- Alcohol consumption, lack of exercise, poor diet and smoking are among the factors that increase the risk of heart disease and cancer.
- While smoking rates are in decline, excessive alcohol consumption is a feature of the middle-aged and middle classes.
- Obesity is a national epidemic.

A complex picture

While a broad 'north–south divide' does exist, the reality of the spatial pattern of inequality in the UK is much more complex. Why is this?

- Variations in morbidity across the UK do not necessarily match patterns of mortality, as different diseases have different distributions. For example, Cornwall has a high risk of malignant melanoma (skin cancer) but one of the lowest rates for heart disease.
- The quality of healthcare services across the UK also varies (e.g. between rural and urban areas).

Remember to consider a range of scales (micro, meso and macro) when analysing spatial patterns.

 Sixty second summary

- Significant variation exists between the average life expectancy of different communities within the UK, with people in London and the south-east living longest.
- Occupation (or lack of paid employment) and lifestyle (poor diet, alcohol misuse, smoking and lack of exercise) are key factors that influence the differing health and morbidity of these groups.

 Over to you

Try to remember the pattern of health or morbidity within the UK in terms of what you'd expect to find, but also what surprises you.

You need to know:

- about a range of factors that influence lifestyle, nutrition and access to healthcare
- what the epidemiological transition is.

Student Book
pages 210–11

Comparing mortality and health (morbidity) in HDEs and LDEs

The greatest causes of mortality include:

- *in HDEs*: heart disease, stroke, dementia and Alzheimer's, cancers, lower respiratory infections (e.g. pneumonia), diabetes and kidney diseases
- *in LDEs*: lower respiratory infections, diarrhoea, stroke, heart disease, HIV/AIDS, tuberculosis, malaria, pre-term birth complications and birth trauma.

Even though many chronic illnesses are preventable, we continue to adopt **sedentary** lifestyles and eat inappropriately, against all advice. Figure **1** summarises factors affecting health (or morbidity) in HDEs and LDEs.

HDEs	LDEs
Wealth – fish, fruit and vegetables are more expensive; less available to poorer groups.	*Poverty* – the dominant cause of malnutrition.
Age – perception of health needs vary depending on age group.	*Pregnancy* – inadequate rest during pregnancy and child-rearing leads to poor health of both mother and child.
Gender – women more likely to seek help (check-ups) and choose to take exercise.	*Environment* – drought, desertification, insect infestation, soil erosion and wetland degradation all add to the difficulties of surviving on marginal land (Figure **2**).
Environment – greater risk of depression in urban areas despite easy access to services.	

🔺 **Figure 1** *Factors, often interrelated, that influence health*

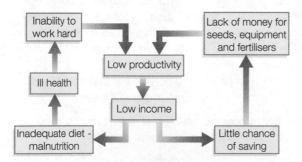

🔺 **Figure 2** *The poorest in LDEs struggle to escape cycles of misery and ill health*

Big idea

Abdel Omram's theory, the *epidemiological transition*, outlines the way in which a population shifts from being defined by *communicable* (infectious) to *non-communicable*, degenerative diseases, with increasing expectancy associated with socio-economic development.

The epidemiological transition

The population dynamics of a country change over time. As the economy of a country develops, so too does the life expectancy of its population. This is the result of investment in improved sanitation, better nutrition and anti-disease programmes. Control of the spread of communicable diseases results in a decline in infant mortality.

However, longer average life expectancy also leads to an increase in the prevalence of chronic diseases of old age, such as heart disease. This change in the types of diseases that shape a population is the **epidemiological transition**. There is evidence of these changes in HDEs during the nineteenth century, and LDEs in the twentieth and twenty-first centuries.

Sixty second summary

- A range of different factors influence health and mortality in LDEs and HDEs.
- A sedentary lifestyle is a key mortality risk factor in HDEs, as is malnutrition in LDEs.
- Evidence of an epidemiological transition can be seen in HDEs and, more recently, in LDEs to a state in which degenerative diseases define the population.

Over to you

Make a list of **three** of the biggest killers in HDEs and then, similarly, in LDEs. Classify **each** disease by type, and state which groups within a population are most likely to be affected.

Student Book
pages 212–13

You need to know:

- the links between location, place and health
- the influence of economic development and environmental factors on well-being.

The city of Glasgow

The '*Glasgow effect*' refers to the health issues and low life expectancy of the population of Glasgow (Figure **1**). This is true even when compared to other UK cities, such as Liverpool and Manchester, which share many similar economic and environmental characteristics. Health campaigns have largely failed to get across their message in Glasgow.

The causes of Glasgow's health problems are complex and inter-related. Deindustrialisation may be the underlying cause (Figure **2**).

 Big idea

Blue Zones are areas of the world where people live longest, on average. Many are remote. The longevity of remote populations demonstrates the importance of *place* – the influence of the environment in which we live, on the health of our communities.

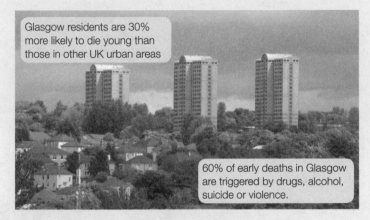

> Glasgow residents are 30% more likely to die young than those in other UK urban areas

> 60% of early deaths in Glasgow are triggered by drugs, alcohol, suicide or violence.

 Figure 1 *Cranhill, a Glasgow housing estate where residents are more likely to die young, compared to those in other UK cities*

Collapse of traditional industries, e.g. shipbuilding

⇩

Areas of Glasgow experience high unemployment rates and have many abandoned industrial sites

⇩

Whole communities lose a sense of identity, pride and 'togetherness' and a collective responsibility for health

⇩

Drugs and alcohol fill the gaps

⇩

Poor mental health and high 'excess' mortality result

 Figure 2 *Why do health and well-being decline as a result of dein-dustrialisation?*

Sri Lanka

Glasgow's health problems may be contrasted with those of Sri Lanka.

- Unlike Glasgow, Sri Lanka's recent health-related targets have been met, as a result of government commitment to disease reduction (e.g. vaccination and health education).
- Reducing the 'disease burden', along with female empowerment, has helped to reduce the birth rate and promote economic development.
- Although Sri Lanka is an LDE, its health provision indicates it is developing economically and moving through Stage 3 of the Demographic Transition Model (see 4.18).

Urban environmental health problems

Scale	Problems affecting the health of urban populations
Within the house	Indoor air pollution from open fires; mould associated with damp
Neighbourhood	Inadequate or polluted water supplies or poor sanitation; discarded drug needles and threats of physical violence
City-wide	Air pollution from traffic congestion; ineffective waste management; river pollution; the urban heat island effect
Extra-urban impacts	Loss of 'green spaces' as a city expands

 Figure 3 *Four scales of urban environmental health problems*

Sixty second summary

- Location or place plays a key role in health and well-being
- Glasgow's population is characterised by large numbers of early deaths, linked to a recent history of deindustrialisation and associated high levels of deprivation – it is a place where health campaigns fail.
- By contrast, the population of LDE Sri Lanka has seen significant health benefits from government commitment to investment in healthcare.

 Over to you

Summarise how the health of Sri Lanka and Glasgow compare in less than **30** words.

You need to know:

- how disease is linked to the physical environment.

Student Book
pages 214–15

Links between disease and the physical environment

In the mid-nineteenth century, John Snow's famous map of a cholera outbreak around the Broad Street water pump in London's Soho, proved the link between contaminated water and disease. The map showed a clear *distance–decay* relationship, i.e. the number of deaths reduced with distance from the pump.

Today, our understanding of the links between our environment and disease is much improved. Nevertheless, our exploitation of resources continues to create new environmental hazards (Figure **1**). Such is the scale of our impact on climate, our current epoch is dubbed the **Anthropocene**.

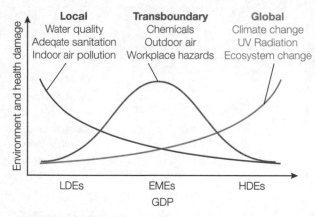

▲ **Figure 1** *Key environmental hazards differ between countries with differing economies*

The impact of the environment on disease

Environmental hazards	Impact
Water, sanitation and hygiene	• Dirty drinking water and poor hygiene result in diarrhoeal diseases • Stagnant water attracts carriers (**vectors**) of disease, e.g. mosquitoes carry malaria
Chemical exposure	• Poisoning from lead piping, asbestos, traces in food
Radiation exposure	• Radon gas may result in lung cancer • Ultraviolet rays may lead to skin cancer
Air quality	• Indoor smoke from cooking causes childhood pneumonia • Smog resulting from industrial or vehicle emissions causes a range of medical conditions, including asthma

▲ **Figure 2**

Look at Figure **1** on page 214 for John Snow's map of cholera outbreak in London in 1854

Addressing environmental health hazards

Despite environmental hazards causing around 25% of all diseases worldwide, less than 5% of global health spending goes on prevention – yet there are significant economic, social and environmental returns on such investment (Figure **3**).

Indoor smoke from solid fuels causes the deaths of around 1.6 million people annually

A socially acceptable solution: flames and smoke are contained, so fewer burns and breathing complications

This low-tech stove burns less fuelwood

▲ **Figure 3** *Clean cookstoves are both sustainable and promote health*

Sixty second summary

- John Snow's map of a cholera outbreak established the link between contaminated water and disease.
- Our understanding of the links between disease and the physical environment has improved over time.
- Environmental hazards include poor air and water quality, exposure to radiation and chemicals like asbestos.
- In the Anthropocene, human influence on climate will influence environmental threats to health.

Over to you

Think of an image to associate with different forms of environmental hazard and/or how such hazards may be addressed to reduce disease. The water pump (with handle removed) on Broad Street may be a good image to begin with.

Student Book
pages 216–17

You need to know:

- the impact of malaria on health and well-being
- the worldwide prevalence, distribution and seasonal incidence of malaria
- the links between malaria and the physical and socio-economic environment.

The disease

Malaria is a biologically transmitted disease caused by the *Plasmodium* parasite, transmitted by the *Anopheles* mosquito. Symptoms include fever, headaches, tiredness, diarrhoea and vomiting, anaemia and jaundice (yellowing of the skin).

Left untreated, malaria causes kidney failure, seizure, coma and death. It can recur up to 50 years after you have been bitten! Malaria may also be caught via blood transfusion, organ transplant, contaminated needles and transfer from pregnant mother to unborn child.

Socio-economic and environmental factors on prevalence and incidence

The severity of malarial infections is affected by the specie of parasite. *Plasmodium falciparum*, most often found in sub-Saharan Africa, causes severe, potentially fatal disease and some are now resistant to drugs. The rate of new cases in an area depends on a number of socio-economic and environmental factors (Figure **1**).

Species	• A long lifespan of the mosquito gives greater time for the parasite to complete its lifecycle inside the mosquito. • Some species of mosquito prefer to bite humans!
Human immunity	• Children are most at risk as they have not developed the immunity their parents benefit from. • Migrant workers or refugees have not been exposed to the disease, so have not built up immunity.
The Environment	• Mosquitoes breed in water, so are abundant in a tropical wet climate. Transmission is most intense during and just after the rains. • The geographic distribution of malaria within countries is complex. For example, there is a risk of infection in rural areas of south-east Asia, where urban areas have been declared disease-free.

⬆ **Figure 1** *Factors affecting incidence of malaria*

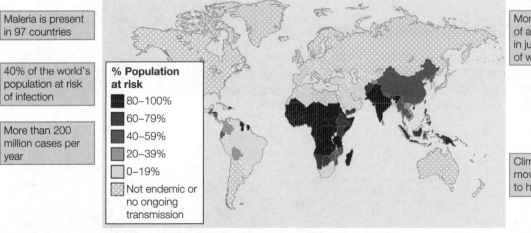

Maleria is present in 97 countries

40% of the world's population at risk of infection

More than 200 million cases per year

% Population at risk
- 80–100%
- 60–79%
- 40–59%
- 20–39%
- 0–19%
- Not endemic or no ongoing transmission

More than threequarters of all malaria deaths are in just 15 countries, 14 of which are in Africa

Climate change is moving the distribution to higher latitudes

⬆ **Figure 2** *Is malaria really a global disease?*

🕐 **Sixty second summary**

- Malaria is a potentially fatal, biologically transmitted disease.
- 40% of the world's population is at risk, though more than 75% of all deaths occur in 15 countries; 14 of these are in Africa.
- Partial immunity is developed over years of exposure, so children and immigrants are most at risk.
- Climate change extends environmental conditions needed for mosquitoes to breed at higher latitudes.

 Over to you

Make the case **for** and **against** the statement 'Malaria is a global killer'.

You need to know:

- about management and mitigation strategies employed to reduce the impact of malaria
- how successful these strategies have been
- the role that international agencies and NGOs have in the fight against malaria.

Student Book
pages 218–21

Malaria in Uganda

In 2015, there were 3.6 million cases of malaria and almost 6000 deaths in Uganda. Why is malaria such a problem?

- High temperatures and rainfall create perfect conditions for mosquitoes throughout the year.
- Large areas of water, such as Lake Kyoga, Albert and Victoria act as breeding grounds for mosquitoes, and are also close to large centres of population (Figure **1**).

What strategies are used to manage or mitigate the impact of the disease?

- Improved drainage and irrigation.
- Insecticides have been successful, though mosquitoes become resistant and so alternatives must be developed.
- Mosquito nets are widely used but aren't always effective as they are often not used properly.
- Early diagnosis and prompt treatment is important in a country where mosquitoes cannot be eliminated.

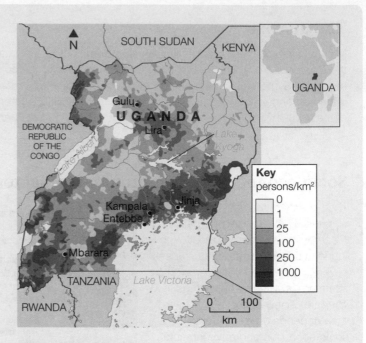

Key
persons/km²

	0
	1
	25
	100
	250
	1000

🔺 **Figure 1** *Population density map of Uganda. How might areas of high population density near Lake Victoria affect the incidence of malaria?*

A Millennium Development Goal met

Malaria's impact has been significantly reduced over the last 15 years (Figure **2**). The Millennium Development Goal to 'have halted by 2015 and begun to reverse the incidence of malaria' has been met convincingly. The WHO estimate that 6.2 billion fewer malaria deaths occurred globally between 2001 and 2015.

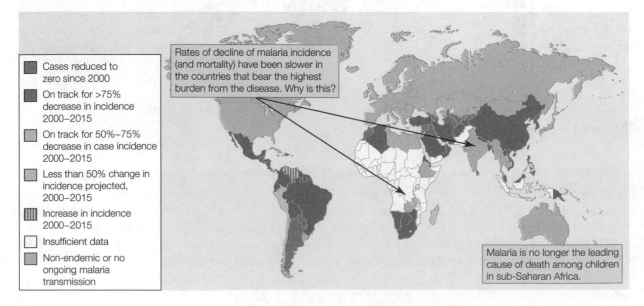

- Cases reduced to zero since 2000
- On track for >75% decrease in incidence 2000–2015
- On track for 50%–75% decrease in case incidence 2000–2015
- Less than 50% change in incidence projected, 2000–2015
- Increase in incidence 2000–2015
- Insufficient data
- Non-endemic or no ongoing malaria transmission

Rates of decline of malaria incidence (and mortality) have been slower in the countries that bear the highest burden from the disease. Why is this?

Malaria is no longer the leading cause of death among children in sub-Saharan Africa.

🔺 **Figure 2** *Headline news: 'Malaria incidence rates projected to fall' (2000–15)*

Management and mitigation strategies that worked

A number of different strategies have been used to reduce the impact of malaria in affected regions.

- Control of the vector using insecticides, even inside dwellings.
- Physical barriers such as insecticide-treated mosquito nets.
- Chemical barriers such as preventative medication for pregnant women; limited use of seasonal anti-malarial drug treatments for children.
- Investing in swift diagnosis – the sooner the disease is diagnosed, the better the chance of recovery.
- Drug treatment in particular highly effective artemisinin-based combination therapies (ACTs).

Despite recent progress, those countries most affected by malaria have seen a rate of decline in the number of cases that lags behind that of other countries. Unsurprisingly, where there is still a high risk of contracting malaria, there is also a weaker health system and lower incomes (Figure **3**).

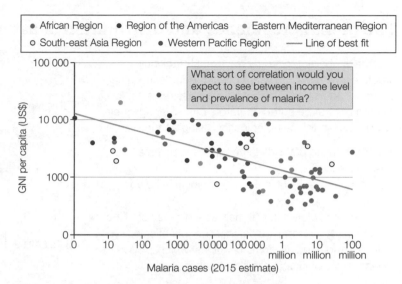

What sort of correlation would you expect to see between income level and prevalence of malaria?

△ **Figure 3** *GNI per capita versus estimated number of malaria cases*

The role of international agencies and NGOs

International agencies (e.g. the WHO) work alongside NGOs to eradicate malaria (the stated aim of Melinda Gates of the *Bill and Melinda Gates Foundation*). These organisations use celebrities to publicise and fundraise for these campaigns (Figure **4**), with the result that:

- between 2005 and 2014, global financing of malaria control programmes increased by 260%
- in 2014, 78% of investment came from international sources (rather than from governments in affected areas)
- in 2016, the UK Government and the *Gates Foundation* pledged £3 billion over five years.

△ **Figure 4** *Former number 1 tennis player Andy Murray wears the* Malaria No More *logo to help publicise the plight of communities affected by the disease and fundraise for the British charity's activities overseas.*

Sixty second summary

- Malaria remains a killer in Uganda where 6000 people died in 2015.
- A combination of physical (climate) and human factors means large numbers of the population are at risk throughout the year.
- At a global scale, the impact of malaria has been significantly reduced and the associated MDG was met.
- A combination of management and mitigation strategies have included using barriers to infection, investment in diagnosis and treatment with highly effective ACTs.
- International agencies and NGOs have been instrumental in funding and facilitating malaria control programmes.

Look at the maps and data in this section (and in the student book) to analyse the spatial pattern of malaria cases at different scales.

Over to you

To help you to revise this topic (as well as 4.13), explain why the rate of the decline in number of cases of malaria varies between different countries.

You need to know:
- the global prevalence and distribution of asthma
- about its links to the physical and socio-economic environments.

*Student Book
pages 222–3*

The global prevalence and distribution of asthma

Asthma is characterised by recurrent attacks of breathlessness, which vary between people in severity and frequency. Though mortality rates are low, asthma cannot be cured. Sufferers must manage their symptoms – often self-medicating using an inhaler.

- An estimated 300 million people have asthma worldwide.
- It is most prevalent in the 10–14 and 75–79 age groups.
- The historic view that asthma was a disease of HDEs (see 4.10) no longer holds; most people affected live in low and middle income countries, which is also where rate of increase in occurrence is fastest.

Mortality rates

Less than 1% of all deaths worldwide are linked to asthma. However the mortality rate linked to asthma varies greatly between countries.

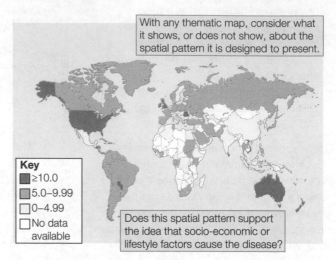

With any thematic map, consider what it shows, or does not show, about the spatial pattern it is designed to present.

Key
- ≥10.0
- 5.0–9.99
- 0–4.99
- No data available

Does this spatial pattern support the idea that socio-economic or lifestyle factors cause the disease?

⬆ **Figure 1** *Percentage of children aged 13–14 suffering from asthma (2009)*

Asthma's triggers

Figure **2** shows a range of environmental and socio-economic triggers for asthma. A family history of the disease is a high risk factor, particularly before the age of 12 years – after which environmental factors are more likely the cause.

Environmental triggers

Exposure to **allergens** that trigger (or exacerbate) asthma symptoms may take place at home or at work (e.g. bakers are at high risk). The time of year is significant – exposure to some allergens varies throughout the year (e.g. tree pollen).

Is a 'Western lifestyle' also to blame?

There is a positive correlation between the increased affluence of a society and the prevalence of asthma. The 'hygiene hypothesis' suggests that, in HDEs, we can afford to live in 'cleaner' homes and lack of exposure to pathogens harms our immune systems.

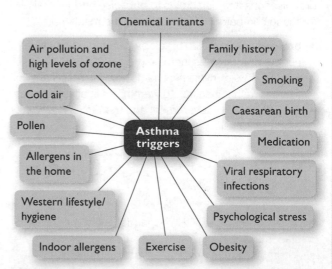

⬆ **Figure 2** *Asthma's triggers are many and varied, some poorly understood*

 Sixty second summary

- Asthma is characterised by attacks of breathlessness and wheezing, affecting 10–14 and 75–79 age groups in particular.
- A disproportionate number of asthma deaths occur in low- and middle-income countries.
- Asthma has many different environmental and socio-economic triggers, including a 'Western lifestyle'.

Over to you

If you know someone who suffers with asthma, ask them to sum up what triggers their symptoms.

Student Book
pages 224–7

You need to know:

- what the impact of asthma is on health and well-being
- management and mitigation strategies.

Impact on health and well-being

Though the *prognosis* (likely course of a condition) is often good for asthma sufferers, at least in HDEs, the day-to-day difficulties of living with the disease should not be underestimated.

Measuring the burden

Because asthma takes effect much earlier in life than other chronic diseases it imposes a higher lifetime burden (Figure 1), not only on sufferers but also on families, carers and the economy.

Dealing with asthma

'I've been dealing with asthma my whole life. I was diagnosed at four and had a few bad asthma attacks. Once I nearly died … My grandad was diagnosed with late-onset asthma in his 60s. So it wasn't a huge surprise when both my sons were diagnosed with asthma too.

Cold and damp weather is the worst trigger for all of us…

…The boys aren't embarrassed to use their inhalers in public. We've always made it a very matter of fact part of life. I try not to let asthma stop us doing anything.' (Adapted from Asthma UK)

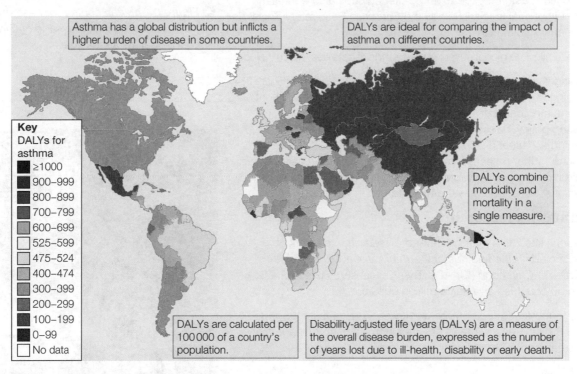

> Asthma has a global distribution but inflicts a higher burden of disease in some countries.

> DALYs are ideal for comparing the impact of asthma on different countries.

Key
DALYs for asthma
- ≥1000
- 900–999
- 800–899
- 700–799
- 600–699
- 525–599
- 475–524
- 400–474
- 300–399
- 200–299
- 100–199
- 0–99
- No data

> DALYs combine morbidity and mortality in a single measure.

> DALYs are calculated per 100000 of a country's population.

> Disability-adjusted life years (DALYs) are a measure of the overall disease burden, expressed as the number of years lost due to ill-health, disability or early death.

🔺 **Figure 1** *Disability-adjusted life years (DALYs) attributed to asthma*

Asthma in the USA

- Ethnic differences in asthma prevalence, morbidity and mortality are highly correlated with poverty, poor urban air quality, indoor allergens and inadequate medical care.
- Higher number of hospital stays amongst African Americans demonstrates the differing impact of asthma by race and community.
- Since the 1980s, asthma has been increasing in the US, in all ages, sex and racial groups.

- The cost to the US economy is huge ($56 billion in 2012) including not only medical costs, but also work absenteeism.
- In 2014, the African American death rate linked to asthma was three times that of their white counterparts.
- A US study of 17000 families estimated that 10.1 million school days are lost across the country in a single year.

Continued over ❯❯❯

Management and mitigation strategies

Once diagnosed, asthmatics can manage their own health in three ways:

- avoid trigger factors
- take preventative medication such as inhaled corticosteroids
- use different prescription drugs to reduce the effects. Salbutamol is an example of such 'rescue relief' and is on the WHO's Model List of Essential Medicines recommended for a basic health system.

If incorrectly diagnosed or if trigger factors are poorly understood, severe asthma attacks can occur, requiring hospitalisation. Inpatient stays are a costly way to manage the disease both for the patient and the wider community.

Role of international agencies and NGOs

International agencies and NGOs are involved in combating asthma on a global scale – the Global Initiative for Asthma (GINA), the Global Asthma Network, the World Allergy Organisation and, of course, the WHO. Scientists, asthma experts and patient groups also work at regional and national levels, such as the European Respiratory Society and the charity Asthma UK.

GINA aims to alleviate global suffering via a range of activities, which include:

- organising *World Asthma Day* to raise awareness of the disease (Figure **3**)
- presenting key recommendations for diagnosis and management of asthma to medical staff
- educating policy makers, to inform investment in treatment
- publishing a *Patient Guide* for sufferers
- promoting further research, by identifying significant areas for further investigation, such as environmental triggers as well as the genetic causes of asthma, and how the two interact.

Disseminating research

Organisations like GINA interrogate and disseminate the work of scientists, including the findings of the *International Study of Asthma and Allergies in Childhood* (ISAAC). The 20-year study used data from more than 50 countries and half a million children.

The study found that three or more weekly servings of fast food were linked to a 39% increased risk of severe asthma among teenagers and a 27% increased risk among younger children.

▲ *Figure 3* 'You can control your asthma' is the positive, enabling message of GINA's annual global awareness day

Figure 4 ▶
ISAAC's findings that link fast food to increased risk of asthma could have global implications

 Sixty second summary

- Asthma takes effect early on in life so imposes a high life-time burden.
- The disease can be managed by using prescription drugs and avoiding of trigger factors.
- In the USA, ethnic differences in prevalence, morbidity and mortality are highly correlated with poverty and poor living conditions.
- Research continues into asthma's triggers as well as its genetic causes. Such work is promoted by international agencies and NGOs.

 Over to you

Pick out **ten** key terms that feature in the *Environment, health and well-being* sub-topic (4.8–4.16) and write your own definitions. Could you use these terms in an exam with confidence?

Student Book
pages 228–9

You need to know:

- about birth and death rates
- infant mortality rates
- fertility and net replacement rates.

Natural population change

- **Natural population change** is the difference between the numbers of births and deaths.
- Net number of migrants is the difference between migrants entering (immigrants) and leaving (emigrants) a country.
- Migration is not included in calculations of natural population change, but is increasingly an important statistic for government planning.

Vital rates of population change

Birth rate

- Expressed as a rate per thousand per year.
- The most common index of fertility.
- Highest in LDEs because of high infant mortality rates and the economic necessity of larger families.

$$\text{Crude birth rate} = \frac{\text{total number of live births in 1 year}}{\text{total mid-year population}} \times 1000$$

Death rate

- The number of deaths per thousand of the population per year
- Important measure as it is the decline in mortality that is largely responsible for population growth.
- Global rates have fallen in the last 100 years as a result of advances in medical care, improved public health and widespread use of vaccinations.

$$\text{Crude death rate} = \frac{\text{total number of deaths in 1 year}}{\text{total mid-year population}} \times 1000$$

Infant mortality rate

- The number of deaths of under one-year-olds expressed per thousand live births per year.
- Generally highest in LDEs (where crude death rates are also highest).

Replacement rate

- Shows the extent to which a population is replacing itself.
- Measured by:
  - *replacement level* – the number of children needed per woman in order to maintain a population size
  - *fertility rate* – average number of children a woman is expected to give birth to in her lifetime
  - *net reproduction rate* – the average number of daughters produced by a woman in her reproductive lifetime (stable population = 1; growing population >1; declining population <1).

Sixty second summary

- Natural population change is calculated by birth rate minus death rate (a negative number suggests population decline).
- Immigration minus emigration is the net number of migrants. This is an important influence on population but is not part of the calculation of natural population change.
- Replacement rate is crudely measured as the difference between births and deaths.

Country	2020–5 Birth rate (per 1000)	2020–5 Death rate (per 1000)	2020–5 Natural increase (%)	2020–5 Infant mortality rate (per 1000)	2020–5 Fertility rate
France	11.4	9.2		2	1.97
Niger	44.5	6.1		49	6.83
Singapore	8.3	5.9		2	1.28
Thailand	9.3	8.8		8	1.41
Zimbabwe	28.6	7.4		38	3.32

🔺 *Figure 1 Natural population change for selected countries (UN, 2017)*

Over to you

In your examination you will need to analyse (examine critically) statistics. Complete Figure **1**, by calculating natural increase. Then, using the evidence, rank order the countries according to level of development.

You need to know:
- about models of natural population change
- their application to countries at different stages of development and in contrasting physical environments.

Student Book
pages 230–1

Thompson's Demographic Transition Model (DTM)

This is the most useful and influential model of natural population change. Thompson classified all countries according to population and wealth, and assumed (unrealistically?) that all would progress through similar 'Western' transitions from simple farming economies (Group C) to complex, modern, urban-industrial ones (Group A).

But, as in all good models, the DTM benefits from having the flexibility to allow changes as circumstances dictate (e.g. the generally accepted addition of a fifth stage, showing declining population in the world's wealthiest Group A countries). There are also suggestions that a sixth stage is needed to account for those countries now experiencing marked net immigration.

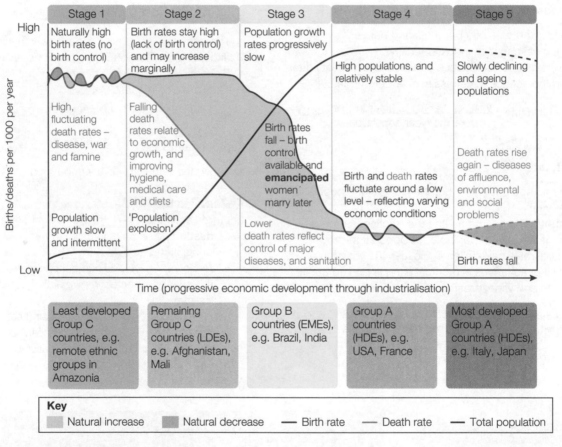

 Figure 1 *Stages in the DTM*

 Sixty second summary

- Thompson classified all countries according to population and wealth.
- The DTM plots the progressive changes in a country's birth rate, death rate and rate of natural increase, in a four-stage path to becoming an HDE.
- A fifth stage has been added to represent further progress towards a state of natural decrease, reflecting ageing populations in HDEs.
- It may be problematic to assume that all countries will develop in the same way.
- Some **demographers** suggest that a sixth stage is needed to account for countries now experiencing marked net immigration.

Over to you

Practise drawing an annotated diagram to describe and explain the five-stage DTM. Then add a sixth stage with annotations to explain, for example, contemporary Germany.

Student Book
pages 232–3

You need to know:
- about population structure
- the dependency ratio.

Population structure

An age–sex (population) pyramid shows the number (or percentage) of males and females in a population, broken down into age groups. Each pyramid shows long-term changes in fertility, mortality and, therefore, growth rates – their shape gradually evolves over time and:

- reflects the social and economic character of individual countries – their state of development
- illustrates specific aspects of the country's demographic history – such as wars and natural disasters
- suggests the type and range of welfare services needed for the population – from maternity care, to preschools through to retirement provision
- shows population trends – and so how to plan for its future.

Dependency ratios

Children and the aged may be considered non-productive and are dependent on the wealth-producing, economically active adults. So, the smaller the adult group relative to the other two, the more difficult it is for a country to be economically viable.

Dependency ratios require careful interpretation because in LDEs, many children work at a young age, whereas in HDEs the reverse is true – children and also many young adults are in education. Furthermore, in LDEs, the aged have always been more productive because of the absence of pensions.

The dependency ratio is calculated using the equation:

$$\text{Dependency ratio} = \frac{\text{dependent population}}{\text{working population}} \times 100$$

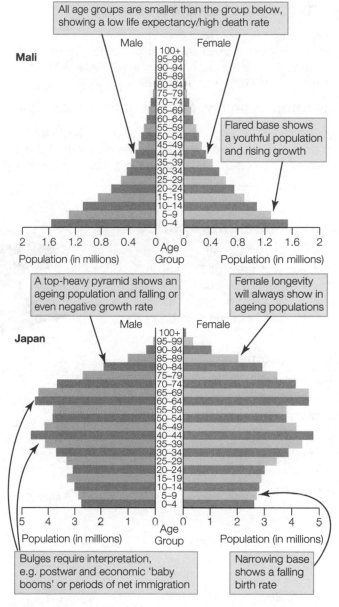

Figure 1 *Age–sex structures for contrasting countries (2014): Mali and Japan*

Sixty second summary

- Population structure is the number (or percentage) of males and females in a population, broken down into age groups.
- It is best shown diagrammatically using age–sex (population) pyramids.
- Distinctive shapes evolve through time, and show different social and economic characteristics.
- The dependency ratio is a calculation of the proportion of a population that is non-productive (children and the aged), and therefore supported by the working age population.
- The higher the dependency ratio, the greater the dependency, and the more difficult it is for a country to be economically viable.

Over to you

Consider likely population structures for countries at different stages of the DTM.

Practise drawing outlines of age–sex pyramid for each, and annotating what they show.

Student Book
pages 234–7

You need to know:
- about cultural controls and social factors that affect natural population change
- some consequences of this population change.

Managing population change

Having a large family in rural communities in LDEs often carries high status – or it did. Today, attitudes are changing as initiatives are taken to reduce infant mortality and to improve maternal health and also the status of women in society. But cultural norms still have a role in influencing birth rates – in some cases they are linked to religious beliefs.

Government policy: enforcement or persuasion?

Over the last 30–40 years, the governments of Bangladesh and China have sought to restrict population growth, but using different strategies. The consequences of these approaches may be appraised, in particular, with regard to the scale of the gender imbalance within these populations today.

China's one-child policy

China has almost one-fifth of the world's population and has gone through an astonishing economic boom in recent decades, much of this 'boom-time' has run parallel with its strict one-child policy for families.

China's problem

In the 1970s, China faced excessive population growth and fear of mass starvation by the end of the twentieth century.

In 1979 the one-child policy was introduced. It worked through incentives and restrictions. It was 'relaxed' in 2013 and, from 2016, replaced by a two-child policy.

The measures were undoubtedly successful, particularly in major urban areas (Figure **1**). But how will history judge China's approach to birth control (Figure **2**)?

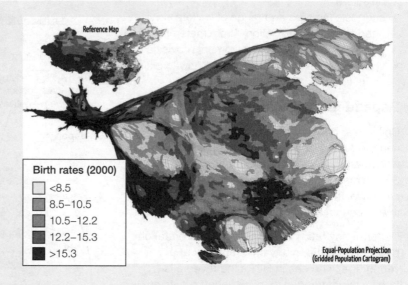

Birth rates (2000)
- ☐ <8.5
- ☐ 8.5–10.5
- ☐ 10.5–12.2
- ☐ 12.2–15.3
- ☐ >15.3

Reference Map

Equal-Population Projection
(Gridded Population Cartogram)

⊙ **Figure 1** *Birth rates in China (2000)*

A low birth rate	A low death rate	A low rate of natural increase	A lower unemployment rate, with production output unaffected
A higher quality of life as demand on social infrastructure and resources is less	Reports of female infanticide, abandonment, forced later terminations and sterilisations	Sex ratio at birth (boys born per 100 girls): in 1982; 108 in 2012; 118	49.9 million 'missing women' (2003 estimate)
An increase in sex trafficking	An increase in prostitution	'Little emperors': a generation of only children	An ageing population with a 'four-two-one problem': one adult supports his or her parents and grandparents.

⊙ **Figure 2** *Consequences of the one-child policy*

Primary health care in Bangladesh

With a population of 170 million, Bangladesh is one of the world's most densely populated countries. Around 80% of the land is low-lying and fertile but is prone to flooding. One in three live in poverty.

Legislation to raise the minimum age of marriage has helped to reduce the fertility rate, but improvements in **primary health care** have been more significant.

Trials in the 1970s showed the value of a doorstep service with trained female health workers, who offered:

- easy access to contraception (and choice) for mothers
- treatment of side effects
- basic maternal and child health care (Figure **3**)

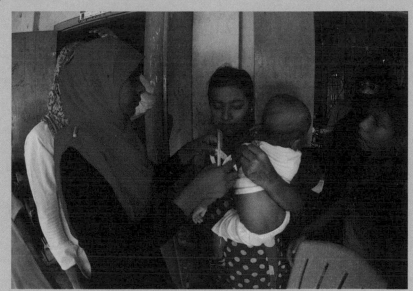

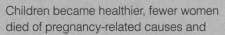 **Figure 3** *Rate of female infant mortality fell fastest in areas where primary health care was provided.*

Children became healthier, fewer women died of pregnancy-related causes and child mortality fell. Families benefited from higher incomes and children stayed in school longer. As a result of this trial, the government invested in training tens of thousands of female primary health care workers.

The consequences

- Bangladesh was one of the first LDEs to meet the MDG of reducing child mortality by two-thirds.
- The population was expected to double by 2050 (to 340 million) but is now likely to reach 200 million before stabilising.
- The rate of infant mortality amongst girls has fallen faster than that of boys – the rate amongst females was previously higher. (Like China, Bangladesh still suffers from a gender imbalance).

'Missing women'

Research has shown that prejudice against females in some countries is not limited to the norms of wider society (e.g. inheritance rights) or the misuse of prenatal screening technology leading to selective terminations.

Such bias extends, often subconsciously, into the family home, by offering females different amounts of food and also varying levels of care during ill health. Globally, the comparative neglect of female children is worse in rural areas and more severe for later-born children within a family.

Calculating excess female mortality

In the 1980s, economist Amartya Sen calculated that globally 'more than 100 million women are missing'.

Subsequent research has supported this claim, noting a 'deterioration [of the situation of women] in China is related largely to strict family planning policies'. (Stephan Kasen and Claudia Wink, 2003).

 Sixty second summary

- China's one-child policy, while successful in reducing population growth, has been criticised for creating problems of gender imbalance.
- Bangladesh's investment in training female primary health care workers reduced the infant mortality rate (particularly in girls), improved maternal health and reduced the fertility rate from six to just over two.
- Around the world, discrimination against girls had led to millions of 'missing women'.

 Over to you

Categorise the consequences of China's one-child policy in Figure **2** into **positive** and **negatives** outcomes. Could you explain how each one links to the one-child policy?

You need to know:

- about the causes, processes and outcomes of migration change
- how these variables affect regions of both origin and destination.

Student Book
pages 238–9

Types of migration

Migration (population movement) can be at all scales from local and regional (in-migration and out-migration) to international (immigration and emigration). Just as births and deaths change the population balance, so also does migration (Figure **1**).

Type of migration	Examples
Temporary – diurnal (daily), seasonal and international	Commuting, **transhumance** and oil exploitation
Step – migration in a series of shorter movements	Farm → village → town → city
Forced – following natural disasters, persecution, conflict or lack of economic opportunity	**Asylum seekers** seeking **refugee** status
Permanent – notably demonstrating *distance decay* (greater distances → fewer migrants)	*Urbanisation* and *emigration* (overseas)
Voluntary – often forming a *migration stream*	Immigration from the Caribbean to Britain following the Second World War

🔺 **Figure 2** *Types of migration*

Push/pull factors

Push and pull factors refer to motivation:

- *Push factors* repel. They range from physical factors (e.g. soil exhaustion) to socio-economic issues (e.g. poverty).
- *Pull factors* attract. They range from a peaceful environment to better job opportunities (e.g. Mexicans seeking the 'American Dream').

Migration outcomes

	Positive	Negative
Origin	• *Overpopulation* pressures eased • *Remittances* support relatives	• Skilled labour shortages • Ageing population structures and gender imbalances
Destination	• Labour pool and diversity increased • Migrants pay taxes and spend money • Cultural and racial variety promotes diversity, integration and understanding	• Housing shortages and pressure on welfare systems • Education and health care services strained • Racial tensions, segregation, crime and violence

🔺 **Figure 3** *Positive and negative outcomes of migration*

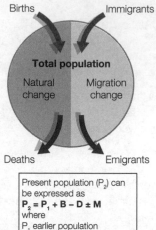

🔺 **Figure 1** *Components of a country's population change; the total population at any one time is the balance between natural change and migration change*

Present population (P₂) can be expressed as
$$P_2 = P_1 + B - D \pm M$$
where
P₁ earlier population
P₂ present population
B births
D deaths
M net migration balance (immigration – emigration)

Sixty second summary

- Migration at all scales can be temporary or permanent, voluntary or forced.
- Forced migrants who flee war or persecution are 'asylum seekers' until officially recognised by the country of destination as 'refugees'.
- Push factors repel people from their home to a new place; pull factors attract people to live and work there.
- There are positive and negative migration outcomes for regions of both origin and destination.

Over to you

Make sure that you understand the jargon associated with migration. Can you instinctively define the following terms: net migration, immigration, emigration, diurnal, transhumance, step migration, refugees, asylum seekers, distance decay, urbanisation, migration stream, overpopulation and remittances?

Student Book
pages 240–1

You need to know:

- about causes, patterns and impacts of international migration from North Africa to Western Europe.

Patterns and causes of international migration

Push factors (poverty, war, persecution) and pull factors (higher paid employment, social care, health provision) frequently combine to cause migration (Figure **1**).

There are three migration routes from North Africa into Europe – the so-called central Mediterranean route to Italy, eastern Mediterranean route via the Balkans and the western route between Morocco and Spain.

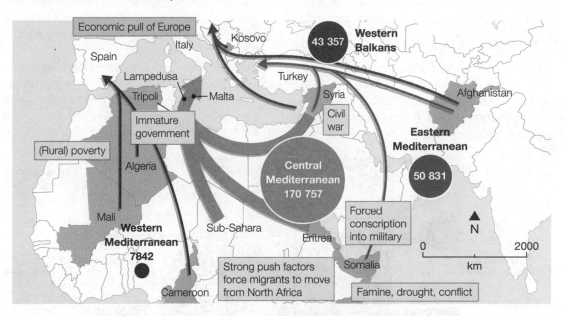

⬆ **Figure 1** *Number of so-called illegal immigrant crossings into Europe (2014) and key causes. Push factors in the green boxes; pull factor in the purple box.*

Impacts on northern and western Africa

- Vulnerable migrants may be exploited by traffickers; growth of organised crime network.
- Potential loss of economically active age groups to the workforce of Europe.
- Human loss – thousands of migrant lives are lost annually when crossing the Mediterranean.
- Remittances, or payments, sent by migrant workers back home contribute to local economy.

Impacts on Italy

- High economic cost of rescue and asylum centres.
- Increasing resentment and public protests towards migrants (e.g. growth of far-right political parties, such as Forza Nuova).
- Pressure on wider international community, including the EU, to tackle root causes of migration and to provide humanitarian assistance.

 Sixty second summary

- Conflict and poverty have pushed millions of migrants from North Africa to Europe.
- Illegal migrants mainly travel along one of three main corridors into Europe; via Spain, Greece/Italy and the Western Balkans.
- Illegal migrants face a range of socio-economic and political challenges.
- Host countries, such as Italy and Greece, have struggled to both process the large numbers of asylum seekers and to provide humanitarian aid, such as food and shelter.

 Over to you

Exam questions will often ask for your opinion. So support any viewpoint with evidence, and aim for a balanced, two-sided answer. For example, since 2015 Germany has accepted more than 2 million refugees – far more than any other nation, including the UK. Do you think that the UK should be doing more to help with the 'refugee crisis'?

You need to know:
- about social, economic, political and environmental implications of migration.

Student Book
pages 242–3

Australia is ...

- one of the richest and most urbanised countries in the world
- a country of enviable agricultural potential and enormous mineral resources – with one of the lowest population densities
- a country of huge environmental challenges, including soil degradation, desertification and habitat destruction.

Population distribution and the environment

Australia's population is mainly concentrated outside the tropics and the hostile semi-desert and desert interior. In fact, in more temperate latitudes, the growing cities are showing signs of **overpopulation** (Figure 1). Although an attractive destination for migrants, immigration brings implications.

Implications of net immigration to Australia

- *Social*: Ethnicity, culture, religion and language have been shaped by young, mainly European families who were encouraged to migrate by the government.
- *Economic*: Since the 1970s, the government encouraged immigration to increase Australia's pool of skilled and professional people. But, in some professions, there is now an oversupply affecting the prospects of young Australians.
- *Political*: Since 2015, the government has hardened its approach to so-called illegal maritime immigrants (Figure **2**), establishing controversial offshore detention centres.
- *Environmental*: See Figure **1**.

Perth's population is increasing.

There have been decades of diminishing rainfall.

One-third of the city's residents were born abroad, most from the UK, New Zealand and South Africa.

City authorities have tried to secure its water supply – plugging leaks, raising awareness of the need for water efficiency.

Dramatic projections have been made about the impact of climate change

Precipitation in SW Australia is forecast to drop by up to 40% by the end of the century.

⌃ *Figure 1 Perth, the capital of Western Australia may be overpopulated*

The Pacific Solution

- Pacific Solution (2001–2008): offshore detention centres were established with subsequent reports of suicide, rape and medical neglect.
- Operation Sovereign Borders (2013): government started to turn back 'illegal smuggler boats'

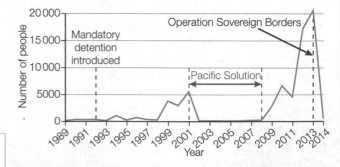

⌃ *Figure 2 Numbers arriving in Australia by unauthorised boat*

Sixty second summary

- Australia has encouraged immigration of skilled professionals as it benefits socially and economically.
- Australian graduates may now suffer from an oversupply of labour and reduced wage rates.
- A changing climate means that Perth's growing population could outstrip available water supplies.
- Housing asylum seekers in detention centres and turning away those who arrive by boat has proved to be controversial.

Over to you

What sort of opportunities and threats does net immigration pose for Australia in the future? Categorise each one under environment, economy, social and political.

You need to know:
- the implications of population size on the balance between population and resources
- the concepts of overpopulation, underpopulation and optimum population.

Student Book
pages 244–5

Population growth on Easter Island

Easter Island is a remote Pacific island, famous for its hundreds of stone statues (Figure 1). It is also infamous for its history of drastic population decline.

- Pessimists suggest that overpopulation of Easter Island resulted in environmental damage (via deforestation), which caused people to run out of food (see Malthus, 4.26).
- Other **anthropologists** propose that rats transported to the island ate the trees, and islanders adapted their diet and agriculture to cope (see Boserup and Simon, 4.26).

▲ *Figure 1 The UNESCO World Heritage site of Easter Island*

Reduced soil erosion

Reduced evapotranspiration

Rock walls protected vegetable crops against wind damage

▲ *Figure 2 Did islanders adapt to survive, using technology to solve environmental problems?*

What is optimum population?

Optimum population is the number of people that can make best use of all available resources within a country or region, so that everyone has an adequate standard of living. Populations above or below the optimum will fail to:

- fully utilise the country or region's resources (**underpopulation**) or,
- support an adequate standard of living for its people, in the short, medium or longer term (**overpopulation**).

Current environmental constraints on population

- *Food productivity* – around 20% more food per person is produced today than 40 years ago but there are still almost 1 billion people who go hungry – those who cannot buy food or have access to land to feed themselves.
- *Water consumption* – freshwater is reducing because of climate change and a growing population (e.g. in south-western Australia). Around 1.1 billion people do not have access to freshwater, mainly in LDEs.
- *Climate change* – with 10% of the world's population living less than 10 m above sea level, the prospect of sea level rise puts homes, crops and livelihoods at risk. Greenhouse gases are being emitted faster than they can be absorbed by shrinking forests and the oceans.
- *Natural hazards* – droughts, floods, wildfires and tropical storms are becoming more frequent, geographically concentrated and on an unprecedented scale. The threat of heatwaves has been given little attention by authorities.

Sixty second summary

- Easter Island was settled 1000 years ago by Polynesians.
- The population peaked at several thousand but was just 700 by 1774.
- The reason for this decline is debated – did overpopulation create an environmental disaster, or did islanders adapt in response to environmental change?
- Easter Island is used as a metaphor for the dynamic relationship between population growth and resources at a global scale.

Over to you

Can you think of a real world example to illustrate each of underpopulation, overpopulation and optimum population?

You need to know:

- about the concepts of carrying capacity, ecological footprint and the demographic dividend.

Student Book pages 246–7

Carrying capacity

Carrying capacity means the maximum number of people that a given environment can support over a sustained period of time.

The average consumption level needs to be decided before modelling carrying capacity. Should we use data on current levels of *Western* consumption (in high-impact HDEs), or include average levels in *low-impact* LDEs in our calculations?

Ecological footprint

Ecological footprint is a measure of the human demands we place on the ecosystems that support us. It is expressed in terms of the amount of biologically productive land needed to:

- produce the resources we consume
- absorb the waste we generate.

Such calculations highlight the extent to which we need to make changes if we want to create a sustainable future.

Demographic dividend

At a *national scale*, South Korea is experiencing a **demographic dividend** due to a falling birth rate in the mid-1960s, and the reallocation of funds from primary education to higher-level education.

This means that there are relatively more people in the economically active age groups. South Korea's economy is booming.

At a *global scale*, there is potential for a demographic dividend because:

- one-quarter of the world's population is aged between 10 and 24 years
- birth rates in many parts of the world are falling.

Those young dependents will soon be working, which may result in faster economic growth and fewer burdens on families.

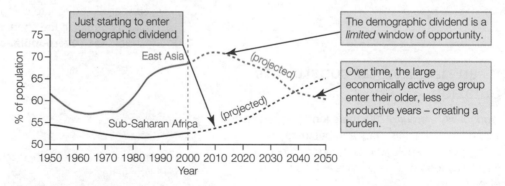

Figure 1 *Working age population (15–59) as a percentage of total population in east Asia and sub-Saharan Africa*

 Sixty second summary

- Carrying capacity is the maximum number of people that a given environment can support over a sustained period.
- The ecological footprint is a way of calculating the impact of human demands on ecosystems.
- Consumption levels, as well as the resources needed to absorb our waste, differ radically between HDEs and LDEs. People in LDEs aspire to HDE living standards.
- South Korea (formerly an LDE) has rapid economic growth as a result of a greater proportion of the population being of working age.

 Over to you

If you were to design your own ecological footprint calculator, what **ten** questions would you pose to find out about the demands each respondent places on the environment?

Student Book
pages 248–9

You need to know:

- about contrasting perspectives on population growth and their implications
- Malthusian, neo-Malthusian predictions and alternatives, such as those associated with Boserup and Simon.

Predictions of future global population

The United Nations produces alternative scenarios for the future global population based on demographic surveys which are projected into the future to produce low, medium and high estimates of population. All scenarios show the population growing until 2050 (Figure **1**).

Pessimistic predictions

In 1798, the English demographer Thomas Malthus gloomily forecast that, unless population growth was slowed by, for example, later marriages, the exponential rise would outstrip food supply and lead to disastrous 'checks' by famine, war and disease.

In *The Population Bomb* (1968) Paul Ehrlich, an American biologist, introduced the suggestion that famines, civil wars and environmental catastrophe indicated a finite future for a planet in crisis.

Optimistic predictions

In the 1960s, the Danish economist Ester Boserup, took a more optimistic view by arguing that agricultural innovations proved that population growth stimulated the changes necessary to support the extra numbers. Using research in LDEs she demonstrated that 'necessity is the mother of invention'.

In *The Ultimate Resource* (1981), Julian Simon suggested that, at times of scarcity, the greatest gains could be made by entrepreneurs substituting new resources and innovating.

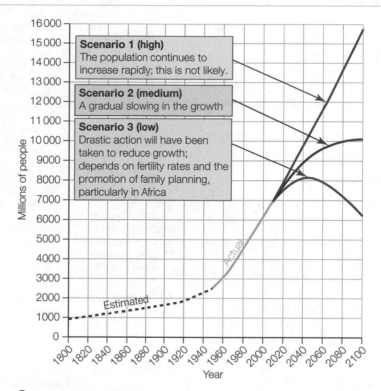

Scenario 1 (high)
The population continues to increase rapidly; this is not likely.

Scenario 2 (medium)
A gradual slowing in the growth

Scenario 3 (low)
Drastic action will have been taken to reduce growth; depends on fertility rates and the promotion of family planning, particularly in Africa

⬤ *Figure 1* UN population estimates 1800–2100

A circular economy

The traditional economy works on the principle of producing, using and disposing. A *circular economy* contrasts with a *linear economy* (make, use, dispose), by keeping resources in use for as long as possible. At the end of their serviceable life, products such as car engines, carpets and washing machines are then recovered and regenerated.

 Sixty second summary

- Concern over rates of population growth has led to attempts to predict future populations.
- All models predict growth until 2050.
- They depend on whether population growth can stimulate innovation or bring about disastrous 'checks'.
- Malthus and neo-Malthusian thinkers, such as Paul Ehrlich and the Club of Rome, were pessimistic about the limits to growth.
- Boserup and Simon suggested times of scarcity bring about innovation – 'necessity is the mother of invention'.

Over to you

Which set of thinkers will be proved right in years to come, the optimists or the pessimists? Argue the case both ways in **100** words.

You need to know:

- how ozone depletion and climate change impacts health.

Student Book
pages 250–3

Ozone depletion

Increased depletion of the stratospheric ozone layer by chlorofluorocarbons (CFCs) has led to an increase in solar UV radiation at the Earth's surface. Impacts of ozone depletion include increased risk of skin cancer, cataracts, reduction of crop yields and a decrease in oceanic plankton (disrupting marine ecosystems).

Skin cancer

Skin cancer is one of the most common forms of cancer. Risk factors include:

- a history of sunburn, particularly in childhood
- outdoor working increasing exposure to the sun
- being fair skinned (more likely to burn)
- old age and a family history of skin cancer.

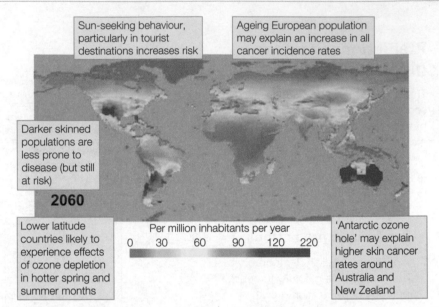

Sun-seeking behaviour, particularly in tourist destinations increases risk

Ageing European population may explain an increase in all cancer incidence rates

Darker skinned populations are less prone to disease (but still at risk)

2060

Lower latitude countries likely to experience effects of ozone depletion in hotter spring and summer months

Per million inhabitants per year

0   30   60   90   120   220

'Antarctic ozone hole' may explain higher skin cancer rates around Australia and New Zealand

⬥ **Figure 1** *Estimate of numbers of extra skin cancer cases by 2060 related to UV radiation (per million inhabitants per year)*

Cataracts

- A slowly developing inflammation of the eye.
- Cause of blindness in 12–15 million people worldwide annually.
- 1 in 5 likely to be caused or enhanced by UV radiation.

Tackling the ozone issue

- Hole in ozone layer above Antarctica discovered in 1985.
- Montreal Protocol (1987) banned use of CFCs.
- Long-term health impacts of thinning of the ozone layer.

Climate change

The 2016 Paris Agreement aimed to reduce carbon emissions so that global warming is restricted to a temperature increase of 2 °C above pre-industrial levels. However, in June 2017, the US announced its withdrawal from the agreement.

Between 2030 and 2050, climate change is expected to cause around 250 000 additional deaths – LDEs are expected to be disproportionately affected (Figure **2**).

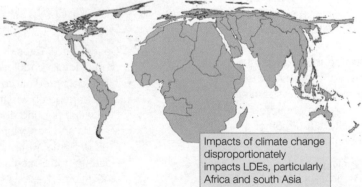

Possible localised benefits, such as decline in winter mortality rates and gains in agricultural productivity, in mid- to high-latitudes

Impacts of climate change disproportionately impacts LDEs, particularly Africa and south Asia

⬥ **Figure 2** *World map drawn proportionally to reflect mortality related to climate change.*

Thermal stress

- Heatwaves likely to increase in frequency and intensity, leading to increased mortality and morbidity rates.
- Greatest health impact on elderly and those with pre-existing conditions.
- Populations in temperate climates, e.g. Europe, are at most risk, as they are less likely to have coping strategies.

- Mortality rates likely to be higher in LDEs as have fewer resources (e.g. air conditioning) to adapt to increased temperatures.
- Decreases in winter mortality may occur, but unlikely that countries will experience an overall net reduction in annual mortality.

Agricultural productivity

	Impacts	Effects
Direct impacts	Increased temperatures	Higher yields in mid- and high-latitudes. Risk of crop failures in seasonally arid and tropical regions.
	Increase in extreme weather events	Drought leads to reduction in crop yields. Heavy rainfall and flooding causes lower quality and yield of crops. Tropical storms negatively impact agricultural systems, particularly in LDEs.
Indirect impacts	Increase in pests and diseases	Pests (e.g. aphids, weevil larvae) more abundant. Changes in rainfall patterns affect migration patterns (e.g. locusts).
	Increased water extraction for irrigation	Increased upstream water extraction may limit water availability for irrigation downstream. Possible political consequences.
	Mean sea level rise	Coastal agriculture flooded and/or affected by salinisation.
Non-climate impacts	CO_2 fertilisation	Increased CO_2 concentration leads to increased photosynthesis and so some crop yields may increase.
	Air pollution (ozone emissions)	Reduces rates of photosynthesis which, in turn reduces crop yield.

 Figure 3 *Impacts of climate change on agricultural productivity*

Nutritional standards

- Changes in patterns of agricultural production would result in food being grown in different locations and by different methods. It will, therefore, have a different vitamin and nutritional content.
- Consumers may choose cheaper, fattier and processed foods if climate change forces the price of 'fresh' food to rise.
- Public health organisations, such as the UK Food Standards Agency, are monitoring and responding to nutritional changes.

Vector-borne diseases

- More than half of the world's population is at risk.
- Changes in temperature affect survival and reproduction rates of vectors.
- Precipitation changes affect aquatic breeding grounds (e.g. mosquitoes).
- Changes in humidity influence vectors such as ticks or sandflies.
- Forced changes to human systems (e.g. increased levels of irrigation), may stimulate vector-borne diseases.

 Sixty second summary

- Depletion of the stratospheric ozone layer by CFCs has led to an increase in solar UV radiation at the Earth's surface, increasing the risk of skin cancer and eye diseases.
- Impacts of climate change on health are likely to disproportionately impact LDEs.
- Patterns of agricultural production may need to change leading to concerns over price and nutritional content of food.
- Expanding populations may come into increased contact with vector-borne diseases as a result of climate change.

Over to you

Climate change is often quoted in answers but remember to be specific in your response and avoid generalisations.

Review the effects of climate change on specific aspects of our health in named regions of the planet.

You need to know:
- the prospects for the global population and projected distribution
- factors in future population–environment relationships.

Student Book pages 254–5

Prospects for the global population and its distribution

Much progress has been made to reduce the global population growth rate, not least regarding the empowerment of women (see 4.29). But the UN still predicts rapid growth in regions of youthful populations, and the potential for shrinkage where ageing populations and low fertility exist (Figure 1).

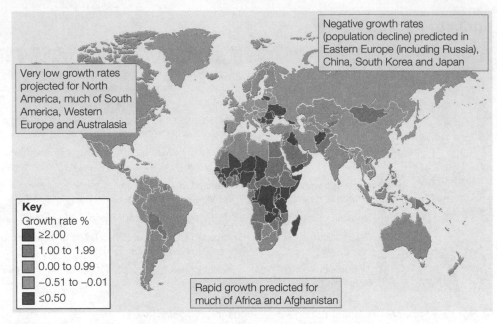

Very low growth rates projected for North America, much of South America, Western Europe and Australasia

Negative growth rates (population decline) predicted in Eastern Europe (including Russia), China, South Korea and Japan

Key
Growth rate %
- ≥2.00
- 1.00 to 1.99
- 0.00 to 0.99
- −0.51 to −0.01
- ≤0.50

Rapid growth predicted for much of Africa and Afghanistan

Figure 1 *Projected annual population growth rate (2010–2050)*

Factors in future population–environment relationships

Climate change: global warming is changing climates, weather patterns and sea levels. Replenishable meltwater supplies are threatened and sea level rise threatens low-lying regions.

Soil erosion: nearly half of all soils are degraded by waterlogging, leaching, salinisation and acidification. Furthermore, erosion is between 10 and 40 times the rate of soil formation (see 4.6).

Destruction of natural habitats: within the next half century another quarter of the remaining forests on Earth will be converted to other uses, such as roads and urban areas.

Loss of wild foods: about 2 billion people, mostly in LDEs, depend on the oceans for protein. Overfishing has resulted in the decline or collapse of valuable fisheries. Effective management could avoid this.

Loss of biodiversity: a significant proportion of the world's wild species and genetic diversity has already been lost; climate change threatens more.

Fossil fuels: still dominate and remain both highly polluting and a significant contributor to global warming. They are also finite (see 5.12).

Unsafe water supplies: most of the world's freshwater is already being used. Underground **aquifers** are also being depleted more rapidly than they are being replenished (see 5.5 and 5.6).

Toxic chemicals: toxins from industry are both manufactured and released as effluent. Health problems have been associated with exposure to toxic chemicals such as in plastics.

Figure 2

 Sixty second summary

- Global population growth rates are slowing overall, but rates are not uniform.
- Rapid growth will continue across much of Africa, but Eastern Europe could be a net loser.
- A number of factors will influence future population–environment relationships, including climate change, the destruction of natural habitats, loss of biodiversity and wild foods, soil erosion, and exposure to toxic chemicals.

 Over to you

Produce a **three** sentence summary of what you have learnt from this topic.

You need to know:

- the character, scale and patterns of population change in Iran
- environmental and socio-economic factors influencing population change in Iran.

Student Book
pages 256–9

The birth of a new state

The modern history of Iran highlights the relationship between the economy, political change and population policy:

1979	Revolution brings rule of autocratic US-backed Shah to end. US sanctions begin
1980	Start of Iran–Iraq conflict; increased numbers of refugees
1980s	Increasing tensions with the US and the West. Economy struggles but health care services improve
1995	Further oil and trade sanctions imposed. Increasing success of national family planning programme
2006	Increased sanctions as result of failure to suspend nuclear programme
2010	Government payments reintroduced for each new child
2012	Abolition of Health Ministry's population control budget
2013	Reformist-backed Hassan Rouhani elected President. Thawing of tensions with the west (and particularly the US)
2016	Part-lifting of sanctions. HDI value (0.766) places Iran in high human development category

▲ **Figure 1** *Iran timeline*

▲ **Figure 2** *Location of Iran*

Population growth

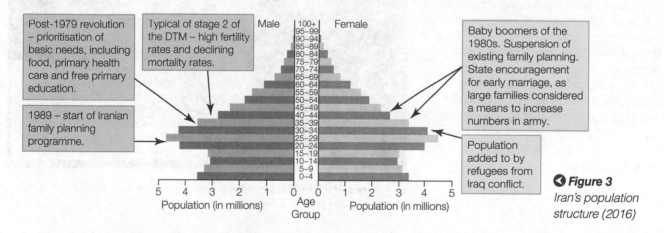

Post-1979 revolution – prioritisation of basic needs, including food, primary health care and free primary education.

1989 – start of Iranian family planning programme.

Typical of stage 2 of the DTM – high fertility rates and declining mortality rates.

Baby boomers of the 1980s. Suspension of existing family planning. State encouragement for early marriage, as large families considered a means to increase numbers in army.

Population added to by refugees from Iraq conflict.

◄ **Figure 3** *Iran's population structure (2016)*

Iranian family planning programme

The threat of hunger (a Malthusian crisis) caused by sanctions and economic costs of conflict prompted/forced the government to introduce a family planning programme as part of the National Five Year Socio-economic Development Plan.

The programme led to a marked reduction in fertility rates (Figure **4**). Its success was the result of a combination of factors:

- Religious leaders gave permission for contraception and male sterilisation.
- Integrated family planning services throughout health care network.
- Improved living conditions – including clean drinking water, electricity, education and health care.
- Role of women – encouragement of female empowerment.
- Public education campaigns in schools, mosques and even in military training.

Continued over >>>

Case Study

An ageing population

- Life expectancy is 76 years.
- Proportion of population aged 60+ is expected to be 12.3% in 2025, more than double the rate in 1966.
- Differences in fertility and migration levels combine to create spatial differences in the elderly.
- Increasing proportion of aged living alone, especially women.
- Government needs to address inadequate pensions.
- Over one-third of over 65s are still working.
- Increasing health burden of the aged.
- Diminishing impact of population dividend – as baby boomers become older, the supply of (cheaper) labour becomes less.

A second baby boom

To counter the effects of the 'ticking time bomb' of an ageing population, the government has again adopted pro-natalist policies. This includes religious leaders and state broadcasts advocating larger family sizes and financial incentives offered for new births.

However, this approach is not proving as straightforward as it was in the 1980s. This is because:

- Iran has developed economically and now ranks 69th out of 188 on the HDI.
- Larger family sizes are no longer considered an economic necessity.
- Younger couples prefer to marry later and continue their education at college and university.
- Female emancipation and more liberal attitudes are allowing women to pursue careers and to gain higher salaries (rather than having families).
- Total fertility rate is currently stable at below two children.

Core components of the Iranian family planning programme

- Freedom to choose contraception method
- Education on population issues
- Respectful of religious and cultural values
- Spacing of pregnancies of between 3–4 years
- Preventing high-risk pregnancies of women under 18 and over 35 years
- Free voluntary sterilisation
- Baby-friendly hospitals

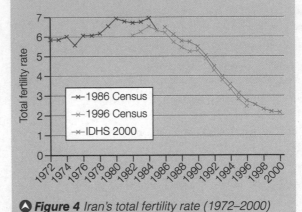

 Figure 4 *Iran's total fertility rate (1972–2000)*

Young women, benefitting from liberal attitudes and empowerment, increasingly aspire to western lifestyles and smaller family sizes. Total fertility rates, at less than two births, are now similar to those of HDEs.

Figure 5 *Western influences are now stronger in Iran than at any time before the 1979 revolution*

 Sixty second summary

- Between 1976 and 1986, Iran's population grew dramatically as a result of encouragement for early marriage and larger families.
- Since 1989, a successful family planning programme and population control measures, such as improved living conditions, education and female empowerment helped to significantly lower the fertility rate.
- Concerns over the impacts of an increasingly ageing population encouraged the government to again adopt pro-natalist policies.
- An increasingly Westernised (and globalised) younger generation no longer see the economic benefits of, nor the necessity for, larger family sizes.

Over to you

Population pyramids such as Figure 3 are invaluable tools for both analysing past and predicting future changes in population structure.

Practise this skill by sketching a population pyramid for Iran in 2046. Then annotate it to show any changes from 2016.

Student Book
pages 260–3

You need to know:
* the relationship between place and health in Abbey Ward, Lincoln.

Case Study

Cities of contrasts

Hull

Hull was the 2017 UK City of Culture. The city council wanted not only to attract multi-million pound investment into the city, but also to change the long-held negative perceptions of the city.

Cynics might argue that an award alone is papering over more systemic cracks but changing attitudes is of crucial importance in bringing about positive change to an area (see 2.6).

Lincoln

Lincoln is an historic city and is not without its own urban problems. The city is home to some of the most deprived areas in England and average earnings of residents are consistently lower than the national average. Furthermore, when specific streets and neighbourhoods are examined in detail, acute levels of deprivation are identified.

Monks Road (Abbey Ward), Lincoln

Within 1 km of the CBD, the Monks Road area of Lincoln is characterised by mixed land use and its population suffers from multiple aspects of deprivation (Figure **2**).

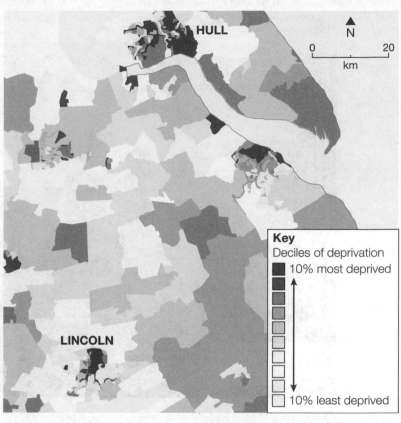

HULL

Key
Deciles of deprivation
■ 10% most deprived

□ 10% least deprived

LINCOLN

⬥ **Figure 1** *Index of multiple deprivation for Hull/Lincoln area*

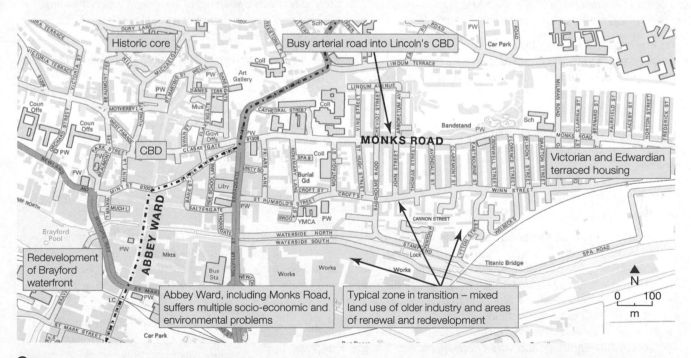

Historic core

Busy arterial road into Lincoln's CBD

CBD

ABBEY WARD

MONKS ROAD

Victorian and Edwardian terraced housing

Redevelopment of Brayford waterfront

Abbey Ward, including Monks Road, suffers multiple socio-economic and environmental problems

Typical zone in transition – mixed land use of older industry and areas of renewal and redevelopment

⬥ **Figure 2** *Physical environment of Monks Road*

Continued over ⟩⟩⟩

Population and socio-economic structure

Compared with all wards in England, Abbey Ward (which includes Monks Road) is in the:

- top 10% most deprived for income
- top 1% most deprived for health and disability
- top 10% most deprived for living environment.

Its population is:

- youthful – the nearby college and university are within a short walk
- multicultural – including, since 2004, migrants from eastern Europe (especially Poland).

Higher than average percentage of young adults – including post-16 students and eastern European migrants

Ageing and, in some cases, deteriorating houses. Grants offered to households for localised renewal (e.g. central heating, insulation)

1960s renewal and redevelopment

High percentage of single-occupancy households

Average income significantly less than national average

High-density urban environment

 Figure 3 *Monks Road – an example of an inner city zone in transition*

Health

Government funded schemes, Community First (2011–15) and Abbey Renewal Area (1998–2008) have improved the health and well-being of local residents, including:

- refurbishment of Abbey Sports Court
- improvements to local homes (e.g. central heating)
- support for community groups, such as City Health Walks
- £3m restoration project of the arboretum
- seven-day NHS walk-in centre.

Local organisations, such as Lincoln City Football Club, work in partnership to support and encourage healthy and active lifestyles.

Community and attitudes

Mobile populations, such as migrants and students, are likely to feel less ownership of an area in which they have lived for only a short period of time. Nevertheless, the area has a long history of community involvement:

- Monks Road Working Men's Club – established for over 100 years.
- Monks Road Neighbourhood Initiative – created in 1998 to support local residents.
- Action LN2 – a group of volunteers who work on smaller projects in the Monks Road area, such as running football sessions or leading language and culture swap sessions.

Sixty second summary

- Lincoln, in the East Midlands, is home to some of the most deprived areas of England.
- Abbey Ward, which includes Monks Road, is in the 10% most deprived for income and suffers from multiple aspects of deprivation.
- The population includes a higher than average concentration of young adults and a multicultural society.
- A range of local community initiatives, funded by the government, Lincoln City Football Club, and charities, provide a range of services to the local population to promote better health and well-being.

Over to you

Relevant place-specific information adds the depth and detail required to raise an answer to the higher levels of a mark scheme.

Review this local-scale example of Abbey Ward in Lincoln and then write four or five sentences that illustrate the relationship between place and health.

5 Resource security

Your exam

(AL) *Resource security* is an **optional topic**. You must answer **one** question in Section C of Paper 2: Human geography, from a choice of **three**: *Contemporary urban environments* **or** *Population and the environment* **or** *Resource security*.

Paper 2 carries 120 marks and makes up 40% of your A Level. Section C carries 48 marks.

Specification subject content (specification reference in brackets)

Either tick these boxes as a record of your revision, **or** use them to identify your strengths and weaknesses

Section in Student Book and Revision Guide	1 ☹	2 😐	3 🙂	Key terms you need to understand Complete the key terms (not just the words in bold) as your revision progresses. 5.1 has been started for you.
Resource development *(3.2.5.1)*				
5.1 What is a resource?				*stock and flow resources, measured reserve, natural resource development*
5.2 Resource peak and sustainability				
Natural resource issues *(3.2.5.2)*				
5.3 Global patterns of energy and water				
5.4 Your way or 'mine'?				
Water security *(3.2.5.3)*				
5.5 Water security, demand and stress				
5.6 Relationship of water supply to physical geography				

5.7 How can we increase water supply?			
5.8 Environmental impacts of a major water supply scheme			
5.9 Management of water consumption			
5.10 Water management and sustainability			
5.11 Water conflict			
Energy security *(3.2.5.4)*			
5.12 Sources of energy and the energy mix			
5.13 Energy supply and physical geography			
5.14 Global energy supplies			
5.15 Environmental impacts of resource development			

5.16 Increasing energy supply				
5.17 Energy management				
5.18 The enhanced greenhouse effect				
5.19 Sustainability issues and energy production				
Mineral security (3.2.5.5)				
5.20 Ore mineral security				
Resource futures (3.2.5.6)				
5.21 Resource futures				
Case studies (3.2.5.7)				
5.22 Water crisis in California, USA				
5.23 Oil exploitation in Alaska, USA				

You need to know:

- about resource classification
- natural resource development over time
- the resource frontier.

Classifying natural resources

Natural resources include minerals and water. Resources may be useful (e.g. a source of energy) or **non-utilitarian** (e.g. a landscape). *Stock resources* are non-renewable (e.g. fossil fuels), *flow resources* are renewable (e.g. geothermal power, Figure **1**).

Mineral *stock resource evaluation* distinguishes between resources and reserves:

- *Resources* are estimated – 'inferred' becoming 'possible' once geologists can reasonably expect them to be economically viable.
- *Reserves* are economically mineable – 'indicated' when mining is judged as feasible and 'measured' when proven to be profitable.

> **Big idea**
>
> Natural resources are those parts of the environment that are of value and use to us.

Flow resources are ongoing – either immediately available (such as here) or created at comparable rates to their consumption (e.g. trees for timber or fuelwood)

Figure 1 ▶
A geothermal power station in Iceland; geothermal power generates 25% of Iceland's electricity

Natural resource development over time

Mineral exploration

The distribution of mineral and energy resources is uneven, but more prevalent in the northern hemisphere with its greater land mass. Geologists use both fieldwork and **remote sensing** in discovery and evaluation, e.g. locating minerals found in **cratons**, fold mountain ranges and within river and marine sediments.

The resource frontier

Resource frontiers can be developed once transport links are established – African and Australian railway maps show examples of routes built specifically to transport an isolated mineral ore to a purpose-built port.

Mineral content of the rock	Geological conditions	Accessibility in relation to the markets
Low-grade deposits produce a comparatively high proportion of waste rock. *High-grade* **ores**, in contrast, will be worth exploiting in the most isolated locations and difficult environments.	Minerals found at shallow depths allow opencast extraction (cheaper than shaft mining), e.g. tin extraction from Malaysian **alluvial plains**.	Low-value minerals (in relation to bulk) are strongly influenced by transport costs – hence the importance of bulk carrier ships transporting low-value ores over great distances at relatively low cost.

▲ **Figure 2** *Mineral exploitation depends on several factors*

 Sixty second summary

- Resources can be human (you), natural and non-utilitarian, with a social value – such as a beautiful landscape.
- Mineral stock resource evaluation distinguishes between estimated (inferred and possible) resources and economically mineable (indicated and measurable) reserves.
- Mineral exploitation depends on the mineral content of the rock, geological conditions and accessibility in relation to the markets.
- Once transport links have been established, resource frontiers can be exploited.

 Over to you

In revising this topic, think about how or why resources, whether human or natural, become *useful*.

Student Book
pages 270–3

You need to know:

- about the concept of a resource peak
- sustainable resource development; the Rössing Uranium Mine, Namibia
- Environmental Impact Assessment.

Resource peak

Stock resource production graphs follow a bell-shaped pattern known as 'Hubbert's Curve'. Look at Figure **1**. The top of the bell marks:

- the point of maximum output (roughly half-way through the reserves)
- when stocks are cheapest. Beyond this point a buyers' market becomes a sellers' market of rising prices.

The decline in production that follows may be hastened by more challenging extraction and production processes because the highest quality and most easily obtained product is extracted first.

The challenge for long-term resource management is to predict these peaks before they happen. But this is not easy because:

- it is difficult to determine the size and nature of future resource demand
- resources are considered exploitable under current, not possible future economic and technological conditions.

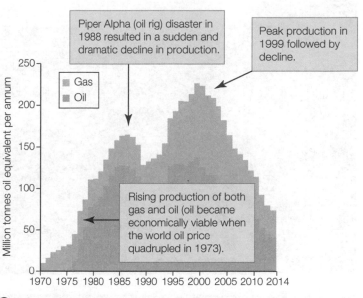

Piper Alpha (oil rig) disaster in 1988 resulted in a sudden and dramatic decline in production.

Peak production in 1999 followed by decline.

Rising production of both gas and oil (oil became economically viable when the world oil price quadrupled in 1973).

⬆ **Figure 1** *North Sea oil and natural gas production 1970–2014*

Sustainable resource development

The word 'sustainable' is often wrongly interpreted as meaning simply environmentally sustainable (e.g. 'unsustainable' fossil fuel exploitation, in terms of depletion of reserves, pollution and climate change). But stock resources can be exploited sustainably – socially, environmentally, economically and politically – by adopting a responsible mining approach. The Rio Tinto Group is a TNC that operates the sustainable Rössing Uranium Mine in Namibia – one of the largest in the world.

Environmental sustainability – met through **stewardship**:
- Crucial water management owing to short supply – with dust controlled by continual spraying.
- All effluent treated and water recycled.
- Leachates controlled and local water sources regularly tested for contamination

Economic sustainability – Rio Tinto Group's financial backing:
- 98% Namibian workforce benefit from full-time contracts with training, career progression, insurance and pensions.
- Rössing Foundation strengthens links with local communities – supports charities, develops education facilities.

Political sustainability – all about mutually assured benefits:
- Revenue generated through licences, **royalties** and taxation.
- The multiplier effect generated which the Namibian government values and nurtures.

Social sustainability – through a health, safety and welfare commitment to its workers:
- Monthly checks for radioactive contamination.
- Building the new town of Arandis with housing supported by health, shopping, educational and recreational services.

◄ **Figure 2**
What makes the Rössing Uranium Mine in Namibia sustainable?

Continued over ▸▸▸

Environmental Impact Assessment (EIA)

This is a process where the environmental impacts of a proposed development or project are evaluated before deciding whether or not to go ahead (Figures **3** and **4**). Any unacceptable impacts can then be reduced or avoided by taking relevant mitigation measures.

 Figure 3 *If a wind farm development was proposed for this area what environmental analysis would precede and guide the decision?*

Mandatory and discretionary procedures to assess environmental impacts	
Description of the project	• Site description, plans for construction, operations and decommissioning • All sources of environmental disturbance (e.g. visual and noise pollution)
Alternatives considered	• Explanation of all alternatives (e.g. other options such as HEP or biomass)
Description of the environment	• Listing aspects of the environment that may be affected (e.g. flora, fauna, soil, water, aesthetic value and cultural heritage)
Description of the significant effects on the environment	• Definition of 'significant' in the context of the project (e.g. noise pollution, collisions with birds)
Mitigation	• Ways to avoid negative impacts should be developed
Non-technical summary	• Preparation of a summary of the Environmental Impact Statement (EIS) • Must be understood by 'the informed lay-person' – no jargon or complicated diagrams
Lack of know-how/technical difficulties	• Final section advising on any areas of weakness in knowledge

Figure 4 *EU directive for completion of an EIA*

 Sixty second summary

- Production of any stock resource will peak at some point in time; diminishing returns will then follow.
- This peak also indicates when stocks are cheapest, beyond which a buyers' market becomes a sellers' market.
- The Rössing Uranium Mine, Namibia shows how stock resources can be exploited sustainably – socially, environmentally, economically and politically.
- An Environmental Impact Assessment (EIA) is the process by which the anticipated environmental effects of a proposed development or project are determined.
- The environmental policy of the EU includes a mixture of mandatory and discretionary procedures to assess environmental impacts.

Over to you

Copy the first **three** bullet points of the Sixty Second Summary adding exemplification to **each one**.

Student Book
pages 274-7

You need to know:

- global patterns of production, consumption and trade/movements of energy
- global patterns of water availability and demand.

Global patterns of production

The world produces more energy today than at any other point in history. The stock resources of coal and oil have traditionally dominated but, since the 1950s, nuclear power and natural gas have become increasingly more important.

Large reserves of stock resources are located in politically unstable parts of the world (Figure **1**). As a result, resource-rich regions such as the Middle East represent a significant geopolitical challenge.

Stock resources are unevenly distributed because of local physical factors, such as climate and landscape differences.

Hydro-electric power (HEP) continues to be the largest contributor to global flow resources.

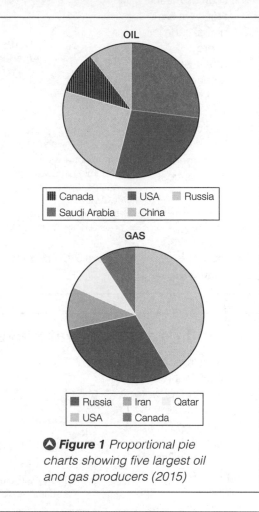

OIL

▥ Canada	■ USA	▨ Russia
■ Saudi Arabia	▨ China	

GAS

■ Russia	■ Iran	▨ Qatar
▨ USA	■ Canada	

Global patterns of consumption

As a global society we are still largely dependent on fossil fuels – they make up around 85% of world energy consumption. This proportion is unlikely to change significantly in the future because rapid industrialisation and increased living standards are increasing energy demand in EMEs.

Nevertheless there remains a significant variation in the energy mix of individual nations and continents.

⬆ **Figure 1** *Proportional pie charts showing five largest oil and gas producers (2015)*

⬇ **Figure 2** *Energy consumption per capita (2014)*

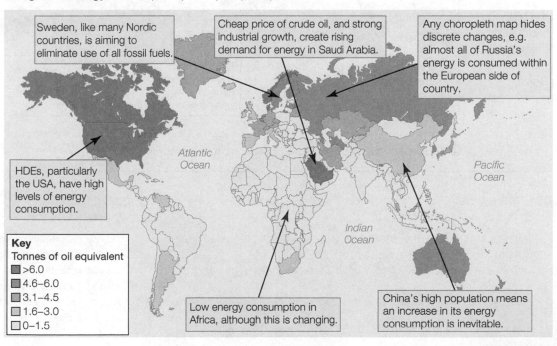

Sweden, like many Nordic countries, is aiming to eliminate use of all fossil fuels.

Cheap price of crude oil, and strong industrial growth, create rising demand for energy in Saudi Arabia.

Any choropleth map hides discrete changes, e.g. almost all of Russia's energy is consumed within the European side of country.

HDEs, particularly the USA, have high levels of energy consumption.

Atlantic Ocean

Pacific Ocean

Indian Ocean

Key
Tonnes of oil equivalent
- ■ >6.0
- ■ 4.6–6.0
- ■ 3.1–4.5
- ☐ 1.6–3.0
- ☐ 0–1.5

Low energy consumption in Africa, although this is changing.

China's high population means an increase in its energy consumption is inevitable.

Continued over ⟫⟫⟫

Global trade and movement of energy

- The unequal distribution of energy sources means that there is a global trade in energy.
- Stock resources may be imported, which is costly, particularly for bulky resources such as coal.
- Many countries are seeking alternative, local energy sources.

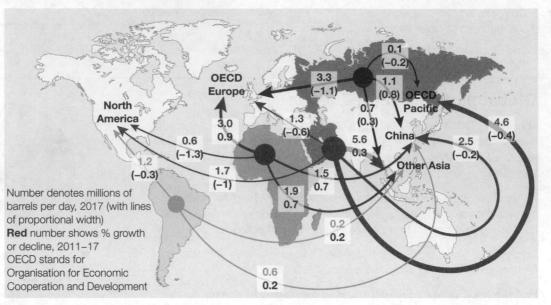

Number denotes millions of barrels per day, 2017 (with lines of proportional width)
Red number shows % growth or decline, 2011–17
OECD stands for Organisation for Economic Cooperation and Development

⬆ **Figure 3** *Crude exports of oil 2017 and growth 2011–17*

Global patterns of water availability and demand

- Even in water abundant regions, demand for water necessitates recycling (see 5.9).
- As population increases, so too does pressure on water resources (Tibetan Plateau, see 5.11).
- Around 1.2 billion people live in areas of physical water scarcity (Murray–Darling Basin, see 5.6).

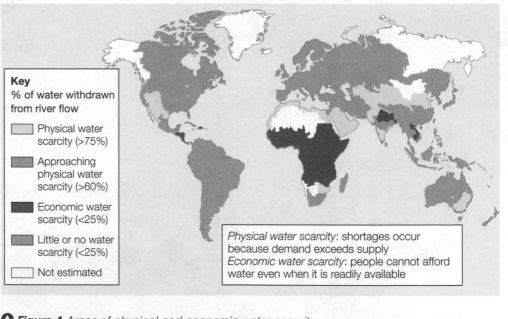

Key
% of water withdrawn from river flow

- Physical water scarcity (>75%)
- Approaching physical water scarcity (>60%)
- Economic water scarcity (<25%)
- Little or no water scarcity (<25%)
- Not estimated

Physical water scarcity: shortages occur because demand exceeds supply
Economic water scarcity: people cannot afford water even when it is readily available

⬆ **Figure 4** *Areas of physical and economic water security*

Sixty second summary

- The last 100 years, particularly the last 50, have seen dramatic increases in global energy production and consumption.
- There is a global shift of production from coal towards oil, natural gas and nuclear power.
- HDEs, particularly in North America, have a far higher energy consumption than LDEs.
- The unequal distribution of energy sources means that there is a global trade in stock resources.
- Around 1.2 billion people live in areas of physical **water scarcity**.

Over to you

You are required to interpret a range of graphs and maps. Choose **one** illustration from Figures **1** to **4** and write **three** geographical statements about the data shown.

You need to know:

- geopolitics of energy
- water resource availability and management.

Student Book
pages 278–81

Energy geopolitics

The success of every successful international power throughout modern history is based on energy resources. Coal, steam and the subsequent Industrial Revolution fuelled the British Empire and oil has helped to establish the USA as a world superpower.

Much has changed since the days of the British Empire, when mining and export of coal, was important. Today, as a result of pit closures in the 1980s, the UK has to import an increasing proportion of its coal.

 Big idea

Geopolitics is the study of ways in which political decisions and processes affect the use of space and resources. It is dynamic as changes might happen quickly, e.g. as government policy shifts as a result of armed conflict.

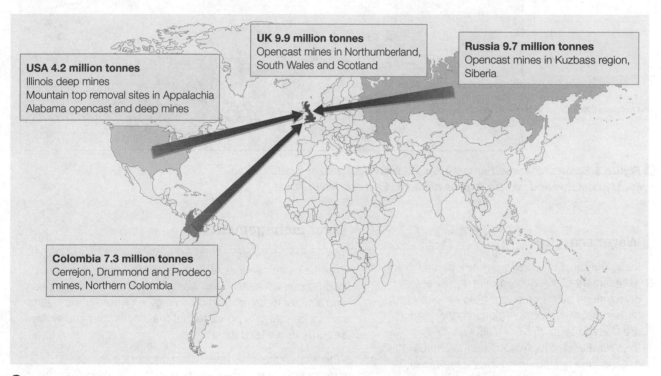

USA 4.2 million tonnes
Illinois deep mines
Mountain top removal sites in Appalachia
Alabama opencast and deep mines

UK 9.9 million tonnes
Opencast mines in Northumberland, South Wales and Scotland

Russia 9.7 million tonnes
Opencast mines in Kuzbass region, Siberia

Colombia 7.3 million tonnes
Cerrejon, Drummond and Prodeco mines, Northern Colombia

▲ **Figure 1** *Where does the coal used in UK power stations come from (2014-15)?*

Alternative resources – shale gas

The so-called 'shale revolution' has environmental and geopolitical implications. Exploitable oil and gas shale reserves are widespread across the globe, including in LDEs, and tend to be closer to centres of population than traditional stock resources.

This raises the possibility of local centres of production meeting local needs rather than being controlled by powerful energy TNCs.

Exploitation of vast shale oil and gas reserves (**fracking**) is likely to help maintain the USA's position as the world's leading geopolitical power.

Figure 2 ◗
Fracking drilling rig in Colorado, USA

Continued over ❯❯❯

Water geopolitics

Almost all of world's largest/greatest rivers are at risk of severe shrinkage and/or pollution. This is a result of human activities and climate change.

Water extraction upstream, will have negative impacts on nations downstream. This is why many trans-boundary river basins demand careful collaborative management.

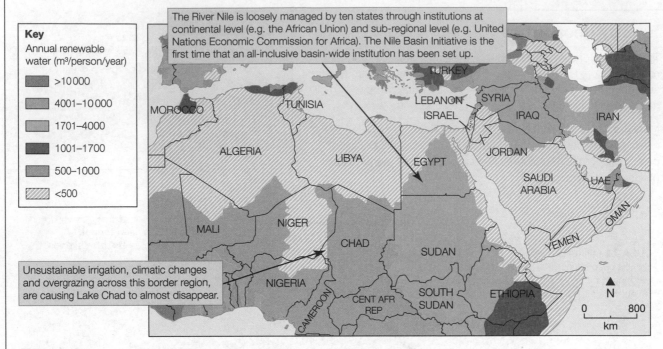

The River Nile is loosely managed by ten states through institutions at continental level (e.g. the African Union) and sub-regional level (e.g. United Nations Economic Commission for Africa). The Nile Basin Initiative is the first time that an all-inclusive basin-wide institution has been set up.

Key
Annual renewable water (m³/person/year)

- >10 000
- 4001–10 000
- 1701–4000
- 1001–1700
- 500–1000
- <500

Unsustainable irrigation, climatic changes and overgrazing across this border region, are causing Lake Chad to almost disappear.

Figure 3 *Some of the world's most insecure water sources are located in the politically unstable and volatile regions of Middle East and North Africa*

Water scarcity

Measurements of water scarcity include an assessment of the demands that resident populations place on supply – in simple terms, higher density (urban) regions in drier regions are likely to be at greater risk of water scarcity than rural and mountainous areas. The US state of California is a contemporary example of an area facing water scarcity (5.22).

River management

The uneven growth in global population will inevitably lead to some settlements increasing in density and extent. Such populations will need clean water for domestic consumption and also for grain production. In regions where basic needs cannot be met, it is possible that fears over food and **water security** could lead to conflict.

Where countries plan ahead and cooperate across international boundaries, such as in Singapore or Senegal, water security can be guaranteed.

 Sixty second summary

- Throughout modern history, the success of every international power has been based upon energy resources.
- The dynamics of energy geopolitics means that power relationships can change quickly.
- Climate change and population growth are creating an increased demand from dwindling water sources.
- Unless countries adopt legal and diplomatic policies towards water usage, dramatic reductions in water supplies could lead to so-called 'water wars'.

 Over to you

Examiners will always appreciate an ability to read beyond the text book. So, carry out further research into a contemporary global example. For instance, wind and solar power in China or the Jordan River basin conflict in the Middle East. Summarise your research in up to **six** bullet points.

You need to know about:

- sources of water, components of water demand and water stress.

Student Book pages 282–3

Sources of water

Although only 2.5% of all the water on the Earth is freshwater, this should still be enough to meet the needs of everyone. Almost all of the freshwater is locked in the ice caps and groundwater (Figure **1**).

Big idea

Water security is the ability of a country to protect access to 'acceptable quality' water resources. At its core, water security is about sustaining peace and political stability.

Water stress

Water stress is when renewable water in a country falls below 1700 m³ per person. It occurs when the demand for water exceeds available water reserves or when poor quality restricts its use.

Water scarcity is when annual water supplies fall below 1000 m³ per person. As the largest centres of population, urban areas are facing the greatest challenges posed by water stress. Consequently, cities are likely to have to use innovative solutions and advanced technologies to meet demand for water.

Water source	Water vol (km³)	% of freshwater	% of total water
Oceans, seas and bays	1338000000	–	96.54
Ice caps, glaciers and permanent snow	24060000	68.6	1.74
Groundwater (total)	23400000	–	1.69
Fresh	10530000	30.1	0.76
Saline	12870000	–	0.93
Soil moisture	16500	0.05	0.001
Ground ice/permafrost	300000	0.86	0.022
Lakes (total)	176400	–	0.013
Fresh	91000	0.26	0.007
Saline	85400	–	0.007
Atmosphere	12900	0.04	0.001
Swamp water	11470	0.03	0.0008
Rivers	2120	0.006	0.0002
Biological water	1120	0.003	0.0001

Figure 1 *Estimate of global sources of water*

Water demand

Figure **2** shows that demand for water is predicted to increase. This is a result of:

- population increase
- urbanisation
- improving standards of living
- increasing demand from more affluent populations and the water-hungry economic activities of emerging LDE and EME economies.

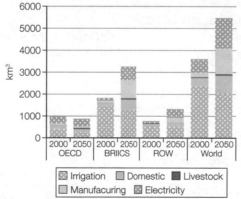

Figure 2 *Global water demand in 2000 and 2050*

Sixty second summary

- As with other natural resources, the Earth's water is distributed unevenly.
- Although 96.5% of all water on Earth is saline, the remaining water sources should be enough to meet our needs.
- The demand for water is predicted to increase by around 55% by 2050 due to population increase, urbanisation and improving standards of living.
- In the future, innovative solutions and more advanced technologies will be required to meet demand, particularly in urban areas.

Over to you

You are frequently required to learn specialist, and sometimes similarly spelt, terms. Write clear definitions of water security, water stress and water scarcity. Creating a glossary for each theme/unit studied is a useful revision tool.

You need to know:

- the relationship of water supply to climate, geology and drainage.

Student Book
pages 284–7

Water supply and physical geography

- *Climate*: lower air pressure is associated with precipitation, however, intense weather systems are likely to result in excessive overland flow rather than the infiltration and recharge of groundwater stores.

  Summer precipitation tends to be less effective at recharging water sources as the ground tends to be harder which discourages infiltration; there is also increased water loss from evapotranspiration.

- *Geology*: permeable and porous rocks, such as chalk, act as aquifers. Natural water gathering basins or **synclines** trap water; **artesian basins** occur when groundwater is trapped between such synclinal layers.

- *Drainage*: river systems are more efficient when there are a range of inputs and a higher drainage density. Overabstraction from groundwater may lead to ecological damage, including, for example, intrusion of saltwater into aquifers at the coast.

The Murray–Darling Basin (MDB)

- The MDB provides 75% of Australia's water and 40% of its farm produce.
- Home to 2 million people but millions more outside MDB, including urban populations of Sydney, Melbourne, Brisbane and Adelaide, depend on it for food and water.
- Average annual rainfall of 480 mm but wide differences in distribution.
- There are additional groundwater inputs into MDB from Great Artesian Basin and Murray Groundwater Basin.
- Outputs from MDB include high evaporation rates in the west.
- Climate is variable and may include extreme weather events, such as flooding and drought, in part a result of El Niño and/or wider climate change. This makes river basin management a challenge.
- MDB is an area of physical water scarcity.
- Managing competing demands, particularly from farmers, is a real challenge.
- Some areas are in a state of almost constant drought.
- Water abstraction upstream has permanently damaged wetland ecosystems near the mouth of Murray River.
- The Murray Darling Basin Authority (2008), tasked with managing the conflicting demands of the MDB, is a key element of the Australian Government's National Plan for Water Security.

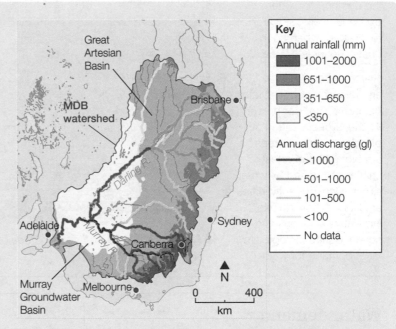

⚑ **Figure 1** *The Murray–Darling Basin; average rainfall and river flow and the two main groundwater sources*

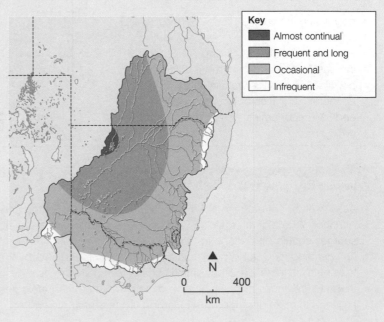

Figure 2 ⊙
Likelihood of drought in the Murray–Darling Basin

Water supply to Greater London

The Thames drainage basin suffers 'severe water stress'. With little spare capacity and a growing population, water has to be managed carefully.

Drainage

- Majority of London's water is taken from River Thames.
- Groundwater is important as most river water is supplied by aquifers.
- Rates of overland flow are higher where tributaries flow over impermeable clay.
- 20% of London's water supply is from groundwater abstraction adjacent to the capital.

Climate

- It is one of driest areas in UK (average annual rainfall is around 25% below national average).
- Higher than average rates of evapotranspiration.
- Two winters of below average rainfall lead to drought conditions.

Geology

- Underlying chalk syncline acts as an aquifer.
- Around one third of water supply is pumped directly from aquifers via boreholes.
- As a result of falling water levels, boreholes are now as deep as 200 m.

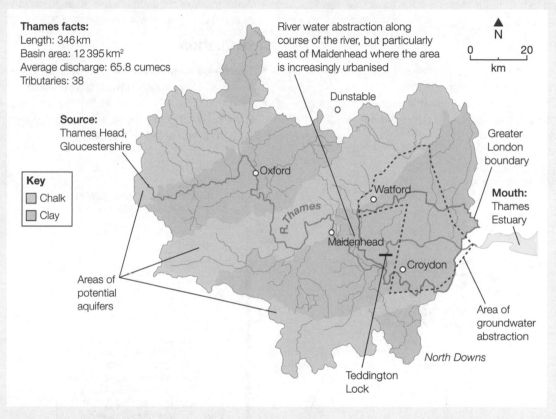

Thames facts:
Length: 346 km
Basin area: 12 395 km²
Average discharge: 65.8 cumecs
Tributaries: 38

River water abstraction along course of the river, but particularly east of Maidenhead where the area is increasingly urbanised

Source:
Thames Head, Gloucestershire

Key
- Chalk
- Clay

Areas of potential aquifers

Dunstable

Oxford

Watford

Maidenhead

R. Thames

Croydon

Teddington Lock

North Downs

Greater London boundary

Mouth:
Thames Estuary

Area of groundwater abstraction

N

0   20
km

 Figure 3 *Water sources and geology of the Thames Basin*

 Sixty second summary

- Access to safe water is seen as a fundamental human right.
- Climate is the major geographical control on water supply.
- River systems that have a higher drainage density and a range of inputs, such as from groundwater flows, are more efficient water gatherers than those that rely on one water source alone.
- Unlike the UK, the Australian Government have a National Plan for Water Security (focused on the Murray–Darling Basin).
- Aquifers are fundamental to Greater London's water supply which relies upon both river water abstraction and also groundwater abstraction. Base flow from aquifers feeds the Thames and its tributaries, whilst one-third of the supply is pumped directly from boreholes.

Over to you

Illustrating a point with a place-specific example transforms an answer. Draw a three column table with headings *climate*, *geology* and *drainage*. Write down place-specific example(s) of how the physical geography of Murray–Darling and/or Thames River Basins influence water supply.

Student Book
pages 288–91

You need to know:

- about strategies to increase water supply, including catchment, diversion, storage and water transfers and desalination.

Strategies to increase water supply

Catchment

- Increased **water catchment** areas – e.g. reafforestation or wetland restoration.
- Diverse management strategies – e.g. encouraging farmers to turn fields from pastoral to arable (reduces surface runoff).

Storage

- Permanent dams – when long-term need and physical geography favours construction.
- Temporary dams – e.g. inflatable rubber dams store water during wet season, and release it for irrigation in drier season.

Diversion and transfer

- Abstraction – water is removed from a natural course, e.g. irrigation, drinking water for urban area (e.g. London).
- Water transfer projects – large-scale to meet increased regional demand, e.g. China (Figure **3**).

Desalination

- **Thermal distillation/reverse osmosis** – converts sea water or brackish groundwater to freshwater.
- Desalination plants – despite high economic and environmental costs, these are increasingly built in drought-prone areas (Figure **1**).

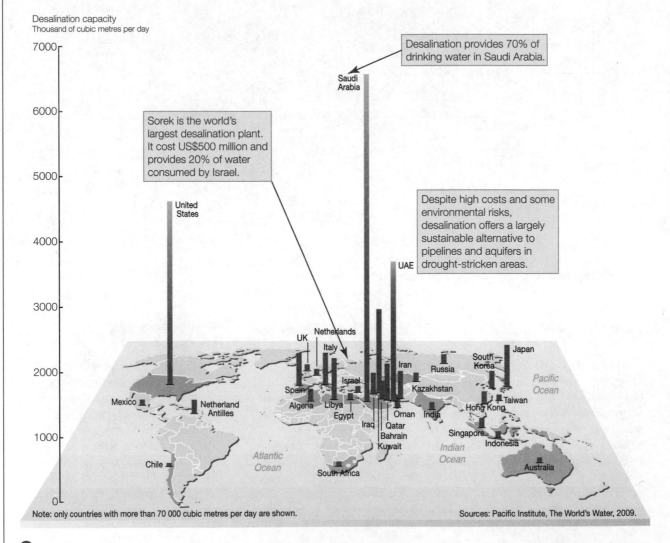

Desalination capacity
Thousand of cubic metres per day

> Desalination provides 70% of drinking water in Saudi Arabia.

> Sorek is the world's largest desalination plant. It cost US$500 million and provides 20% of water consumed by Israel.

> Despite high costs and some environmental risks, desalination offers a largely sustainable alternative to pipelines and aquifers in drought-stricken areas.

Note: only countries with more than 70 000 cubic metres per day are shown.

Sources: Pacific Institute, The World's Water, 2009.

🔺 **Figure 1** *Global pattern of water desalination capacity*

A large-scale water transfer project – China

- China's South–North Water Transfer Project is a **megaproject**.
- It began construction in 2002 with a completion date of 2050.
- The drier north, including megacities such as Beijing, relies on water transfers from the wetter southern regions.
- Water will be transferred to the drought-affected north via 2400 km of tunnels and canals.
- It aims to abstract 45 billion m³ of water from Yangtze River basin per year.
- Reduced discharge of Yangtze River is negatively impacting its ecosystem.
- There are socio-economic costs where residents forced to relocate.
- Huge cost of US$62 billion.

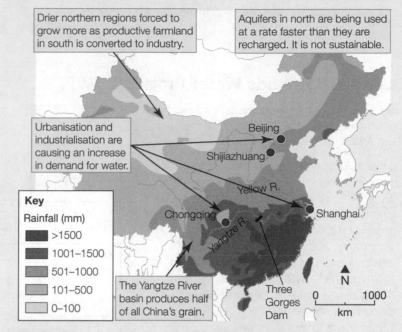

Drier northern regions forced to grow more as productive farmland in south is converted to industry.

Aquifers in north are being used at a rate faster than they are recharged. It is not sustainable.

Urbanisation and industrialisation are causing an increase in demand for water.

Key
Rainfall (mm)
- >1500
- 1001–1500
- 501–1000
- 101–500
- 0–100

The Yangtze River basin produces half of all China's grain.

Figure 2 *China's rainfall is unevenly distributed*

———— Western (under review): If built, this scheme would divert 20 billion m³ of water from three tributaries of the upper Yangtze. Such is the scale and engineering required (300 km of tunnels, 200 m high dams, permafrost, tectonic activity) that it may never be built.

———— Eastern (1155 km): Completed in 2013. Includes two 9.3 m diameter tunnels 70 m under the Yellow River.

———— Central (1267 km): Completed in 2014. The expansion of the reservoir at Danjiangkou has forced around 350 000 people to be resettled.

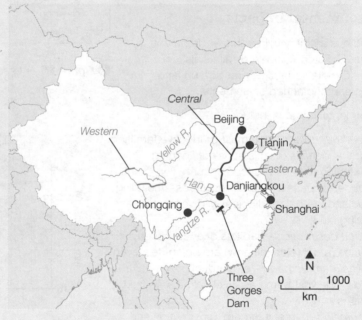

Figure 3 *China's South-North Water Transfer Project*

⏱ **Sixty second summary**

- All over the world, populations are searching for new water supplies.
- Water transfer from areas of excess to areas with a shortage has a long history – dams facilitate water diversion by aqueducts, canals and pipelines.
- Construction of water catchment areas allows more rainwater to be collected, e.g. wetland restoration or reafforestation.
- Desalinisation by thermal distillation or reverse osmosis is controversial, highly expensive and environmentally risky.
- China's ambitious and expensive South–North Water Transfer Project aims to help supply the drought-affected north.

✏ **Over to you**

Write **four** sentences to summarise the strategies typically used to increase water supply. Try and illustrate **each one** with a specific example – e.g. China (5.7), Lesotho (5.8), California (5.22).

You need to know:

- how a major water supply scheme can impact the environment.

Student Book
pages 292–3

Lesotho Highlands Water Project (LHWP)

Lesotho is a poor mountainous country that is surrounded entirely by the much wealthier South Africa.

- The LHWP was signed in 1986 and aims to transfer water from Lesotho to South Africa.
- South Africa pays the water transfer costs and Lesotho finances the hydroelectric power projects (Figure **1**).
- Five dams, 200 km of tunnels and water pumping stations built across several stages make up the LHWP.
- The Maluti Mountains are the source of the Senqu (Orange) River.
- Mountains and mountain wetlands, are an important part of southern Africa's water resources.

Environmental impacts

Lesotho has a highly vulnerable environment as a result of steep slopes; fragile soils; soil erosion; climate variability; pressure on land from farming and unplanned settlements.

Environmental impacts include:

- *Threat to farmland* – the most productive farmland (e.g. the Mahale area) is at risk of flooding, threatening the nation's food security.
- *Reduced access to natural resources* – including woodland, herbal medicines, wild vegetables and grazing land.
- *Habitat destruction* – habitats of endangered species (e.g. Maluti minnow) are being destroyed.
- *Downstream damage* – e.g. reduction in natural flooding downstream negatively affects wetland habitats.
- *Soil erosion* – a result of dam construction and associated infrastructure. Also caused by displaced communities that have been forced to farm more marginal lands on steeper slopes.

Key

- ------ International boundary
- Reservoir
- ■ Dam (completion date)
- — River
- 2500m–3000m elevation
- → Tunnel

0   50 km

 Figure 1 *The Lesotho Highlands Water Project (LHWP)*

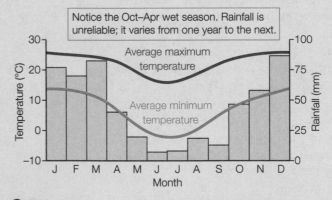

Notice the Oct–Apr wet season. Rainfall is unreliable; it varies from one year to the next.

 Figure 2 *Annual rainfall and temperature graph for Lesotho*

Sixty second summary

- Lesotho is poor, landlocked and mountainous but with an abundance of water.
- The LHWP aims to supply South Africa with water.
- It involves five dams, 200 km of tunnels and water pumping and transfer stations.
- Significant environmental impacts of the project include loss of farmland, reduced access to natural resources, habitat destruction, downstream damage and soil erosion.

Over to you

In no more than **50** words, summarise the environmental impacts of the LHWP.

Student Book
pages 294–5

You need to know:

- how climate change impacts UK water supplies and demand
- about strategies to manage water consumption
- what our water footprint is.

UK Climate Impact Programme predictions

The UK climate is expected to become more extreme, e.g. droughts in south and east, increased risk of significant flood events.

The impacts of climate change on UK water supplies and demand include:

- *Flooding* – rising sea levels, storm surges and high intensity rainfall are forcing the Environment Agency to re-evaluate their flood models.
- *Disease* – drier conditions will lead to populations being placed at increased risk from some insect-borne diseases (e.g. malaria).
- *Heatwaves* – increased demand for water, e.g. for irrigation.

Water footprint

Our water footprint includes a measure of the amount of **virtual water** used to produce each of the goods and services that we use (such as growing food, making our clothes and even our digital gadgets).

 Sixty second summary

- Our water footprint measures the amount of 'hidden' water used to produce the goods and services we use, as well as our direct water usage.
- The UK climate is expected to become more extreme (hotter and drier in summer, milder and wetter in winter) – more storm/flood events, droughts, heatwaves, with associated health risks (skin cancer, insect-borne diseases).
- Our water-intensive lifestyle is, arguably, unsustainable; emerging economies also aspire to a similar extravagant use of water.
- To meet global water needs, any increase in supply must work alongside conservation measures and strategies.

 Over to you

Make a list of strategies that might be used to reduce our water footprint.

 Big idea

Increased demands, as well as the impacts of climate change, are having a significant and long-term impact on the water cycle.

Strategies to manage water consumption

Food consumption

Food retailers could source food from regions that do not suffer water stress. This will mean that out-of-season fruit and vegetables (grown using irrigation) are no longer imported.

Water conservation in the home

- Water meters force us to think about management of water. Water usage reduces by around 10–15% following installation.
- Use water-efficient kitchen devices.
- Install dual flush toilets that use less water for each flush.
- Turn off taps while washing-up or brushing teeth.
- Install tap aerators.
- Use a shut-off nozzles on hosepipes.
- Use plants that need less water and use mulch to retain moisture.
- Use a broom and not a hosepipe to clean drives and footpaths.
- Water gardens early or late in the day to reduce evaporation.
- Install greywater recycling and rainwater harvesting systems.

Activity	Water used
Running the tap	8–12 litres/min
Washing up in the sink	6–8 litres
Washing hands and face	3–9 litres
Taking a normal shower	6–12 litres/min
Taking a power shower	13–22 litres/min
Flushing the toilet	5–12 litres
Running a dishwasher	15 litres
Running a washing machine	60–80 litres
Having a bath	75–90 litres
Using a hosepipe	550–1000 litres/hr
Making food and drink	6–10 litres

▲ **Figure 1** *Lifestyles of HDEs are considered to be increasingly water intensive*

Student Book
pages 296–7

You need to know:
- about issues associated with diffuse water pollution and water management.

Diffuse pollution

Diffuse pollution occurs when small amounts of pollutants, often from many different sources, are washed into a water catchment across a wide area. The effects are more significant than an individual pollutant because the sources are widespread, and so hard to spot.

 Big idea

Sustainable water management addresses the needs of all users while maintaining water quality and flow rates. It involves changing attitudes towards our use of finite water resources.

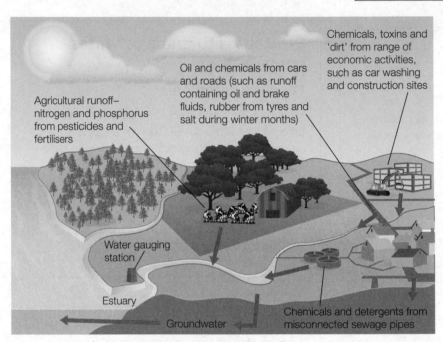

Chemicals, toxins and 'dirt' from range of economic activities, such as car washing and construction sites

Oil and chemicals from cars and roads (such as runoff containing oil and brake fluids, rubber from tyres and salt during winter months)

Agricultural runoff–nitrogen and phosphorus from pesticides and fertilisers

Water gauging station

Estuary

Groundwater

Chemicals and detergents from misconnected sewage pipes

The composting toilet

WaterAid is an NGO that aims to improve access to safe water, sanitation and hygiene for everyone. It has been trialling sustainable composting toilets in Mozambique. The project is sustainable because:

- water is conserved as no water is used in toilet flushing
- pollution of groundwater (an important drinking water supply) is prevented
- it is hygienic and does not attract flies (and so disease).

◀ **Figure 1** Diffuse pollution

Examples of water management

Virtual water trade

As products are traded internationally, their water footprint follows them – a virtual water trade. So, water-scarce countries may import water-intensive goods, rather than producing them domestically.

Greywater recycling

Greywater is domestic 'dirty' water (not sewage) from baths, sinks and household appliances. Greywater recycling systems clean the water for reuse.

Rainwater collection recycling

Rainwater harvesting systems (RHS) divert water from roofs to flush toilets and wash clothes. RHS is cheaper than greywater recycling as it needs less cleaning.

Groundwater management

Groundwater supplies 30% of all available freshwater. Management is therefore important to prevent overabstraction. The Environment Agency charges users for an abstraction licence.

 Sixty second summary

- Diffuse pollution occurs when small amounts of pollution, from agricultural, industrial, transport and domestic sources, are washed into a water catchment across a wide area.
- Greywater recycling, rainwater collection recycling, groundwater management, virtual water trade and conservation (see 5.9) are sustainable approaches to water management.
- Water Aid's composting toilet project succeeds in conserving water, preventing groundwater pollution and changing perceptions towards use of urine as a fertiliser.

 Over to you

Sustainability is arguably an overused and misplaced term that causes confusion. Write your own definition of sustainability. Using examples, explain the term *water sustainability*.

Student Book
pages 298–301

You need to know:

- about water conflicts at local, national and international scales.

The politics of water

Increased competition for water, political instabilities, poor management and the impacts of climate change mean that water conflict is becoming more frequent at local, national and international scales (Figure **1**).

Parties involved	Scale	Violent conflict or in context of violence	Basis of conflict	Description
South Africa	Local	Yes	Development dispute	At least four people killed during protests over water shortages in the northern town of Brits.
Syria	National	Yes	Political and Military tool	Water for the city of Aleppo is cut off; pumping stations and water distribution networks are bombed.
Ukraine and Crimea	International	No	Political tool; Development dispute	After Russia annexed Crimea, Ukraine accused of cutting water supply in the North Crimea Canal, leading to a water shortage for Crimea's agricultural fields.

⊙ **Figure 1** *Examples of water conflicts in 2014 at different geographical scales*

Will the Tibetan Plateau determine Asia's future?

The Tibetan plateau:

- covers an area of 6 million km²
- is hugely significant for both the health of the planet and the geopolitical stability of Asia
- is the source of the Yellow River, which supplies water to 20% of China and serves 50 cities; it has sometimes slowed to a trickle
- is one of the fastest-warming areas on Earth – a potential global environmental catastrophe. Melting permafrost could change the summer monsoon, causing droughts in northern India and in the wheat-growing northern China. More intense and frequent floods (and mudslides) are also likely.

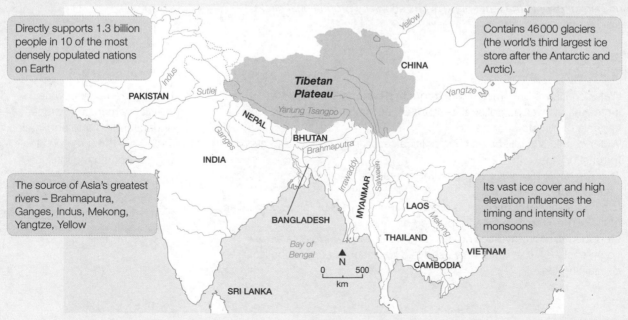

Directly supports 1.3 billion people in 10 of the most densely populated nations on Earth

Contains 46 000 glaciers (the world's third largest ice store after the Antarctic and Arctic).

The source of Asia's greatest rivers – Brahmaputra, Ganges, Indus, Mekong, Yangtze, Yellow

Its vast ice cover and high elevation influences the timing and intensity of monsoons

⊙ **Figure 2** *The geopolitical and environmental importance of the Tibetan Plateau*

Continued over >>>

China's water megaproject

China's current (and future) plans to dam or divert water from five of the large rivers coming out of the Tibetan plateau. This is not only exacerbating environmental problems but also raising tensions in a geopolitically sensitive region of the world (Figure **3**). China's proposed western route of the South-North Water Transfer project (see 5.7) threatens the peace and security of the entire region.

In turn, India is constructing dams and hydro projects on the Brahmaputra River (which has its source in the Tibetan Plateau). In one of the most crowded regions on the planet, the construction of upstream dams, barrages, canals and irrigation systems, demands careful political discussions to avoid a potentially wide-ranging water conflict.

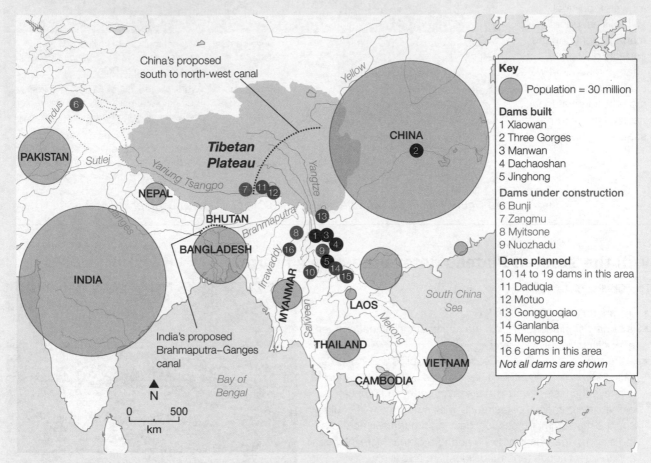

Figure 3 *Turning the taps off – plans to manage the flow of water from the Tibetan Plateau will affect huge numbers of people (national populations shown by proportional circles)*

 Sixty second summary

- Water conflict is becoming more frequent as competition for water increases, as well as political instabilities, poor management and the impacts of climate change.
- Water conflicts at all scales have political, economic, military and even terrorist origins.
- The Tibetan Plateau is a vast, environmentally strategic area critical to the health of the planet.
- It is geopolitically significant in directly supporting 1.3 billion people in ten of the world's most densely populated nations.
- Megadams, micro-hydro, irrigation and water transfer projects across south-east Asia can be openly used in a water conflict or as a political bargaining tool.

 Over to you

Recent examples of water conflict demonstrate higher-level understanding and application. Tables are great revision tools for organising information.

Use Figure **1** as a template to create your own table of water conflicts, which have happened during the current academic year.

You need to know:

- the sources of primary and secondary energy
- about demand and energy mixes in contrasting settings.

Student Book
pages 302–3

Primary and secondary sources of energy

Energy sources are categorised as *stock* or *flow*, and *primary* or *secondary*. Primary sources are raw materials used in their natural form (e.g. coal, petroleum and wood) and often converted into more practical secondary energy sources such as electricity and petrol.

Global demand and the energy mix

The different primary energy sources used to meet demand make up the *energy mix* – this will change over time but:

- it is still dominated by HDEs and fossil fuels
- as LDEs develop demand increases
- increased demand reduces stock and prices rise
- alternative flow resources becoming more significant.

STOCK		
Type	**Key characteristics and uses**	**Key locations and trends**
Coal (Primary)	• Dirty, bulky and difficult to transport • Electricity generation, metal smelting and heavy chemicals	• Declining – older industrial countries • Increasing – Russia, India and China (biggest producer and consumer)
Petroleum (oil – Primary)	• Relatively cheap – transport, industry, electricity generation and petrochemicals • OPEC producers dominate supply	Future potential – • Arctic Ocean and South Atlantic • Oil shale 'fracking' (see 5.4)
Natural gas (Primary)	• Flexible and 'clean' – domestic, industry and electricity generation • Pipeline transport, but some by sea as LPG	• Leading producers – USA, Russia, Canada, China • European production notable but declining (e.g. North Sea) • 'Fracking' potentially abundant
Uranium (Nuclear fission – Primary)	• Expensive, contentious, 'clean' and extraordinarily powerful – for electricity generation • Huge potential in combating climate change	• USA produces the most nuclear energy • EMEs rapidly developing the technology • China alone could build 300 more stations by 2050
FLOW		
Biomass (Primary and secondary)	• Fuelwood – for heating, lighting and cooking • Biogas (biomethane) from organic decomposition to generate electricity • Alcohol (bioethanol) from fermenting crops (e.g. cereals) • Biodiesel from oilseed rape	• Main source of energy in LDE rural areas • 40% of world's trees removed for fuelwood • Ethanol and biodiesel alternatives for transport • Biomass (wood pellets) for electricity generation (e.g. Drax, North Yorkshire)
Hydroelectric power (HEP) (Primary)	• Low operating costs – but high installation and transmission costs • Dams and reservoirs can also control Floods, and supply domestic, industrial and irrigation water	• Important in mountainous HDEs (e.g. Switzerland), but huge potential in Africa (e.g. Lesotho, 5.8) • Multi-purpose large-scale schemes notable in LDEs and EMEs • Tidal power potential (e.g. Severn and Humber in UK)
Other renewables (Secondary)	• Solar, wind, wave, tidal and geothermal energy (see 5.16)	

⬤ *Primary and secondary stock/flow energy sources*

Sixty second summary

- Energy sources are categorised as stock or flow, and primary or secondary.
- The world today sees global demands and an energy mix still dominated by (diminishing) fossil fuels, but alternative flow resources are growing in significance.
- Different primary energy sources are used to satisfy increasing demand; they make up the changing energy mix.

Over to you

Figure **2** on page 303 in the student book projects the anticipated global energy mix up to 2040. Summarise its trends in **one** sentence.

You need to know:

- the relationship of energy supply to geology, climate and drainage
- China's energy demands, mix and security.

Student Book
pages 304–5

Energy and geography

Physical geography can determine a country's energy mix – geology (fossil fuels); climate (solar energy); drainage (HEP).

Human geography is also a factor in achieving **energy security** – how much energy is required and in what forms. It is relevant in quality of life issues, such as the environmental impact.

China's energy demands, mix and security

China is the world's biggest producer and consumer of energy, largest net importer of oil and biggest producer of greenhouse gases.

But China's energy mix is varied (Figure **1**) and will change as social, economic and environmental circumstances dictate. It has a generous range of its own energy resources, but many are in remote locations and not economically viable (Figure **2**).

Over two-thirds of China's electricity is generated by highly-polluting coal-fired power stations.

Despite its resources, such are China's demands, that 90% of the energy it uses has to be imported! China is **energy dependent** rather than energy secure.

Renewable energy in China

China is slowly adding wind, solar and biomass renewables to its energy mix and HEP already accounts for 22% of electricity generation, which raises its own concerns:

- flooding of established settlement, infrastructure and productive agricultural land
- displacement of population
- reservoir silting and pollution
- vulnerability to earthquakes and landslides.

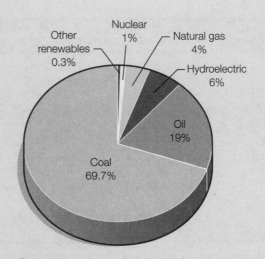

▲ **Figure 1** *China's energy mix (2012)*

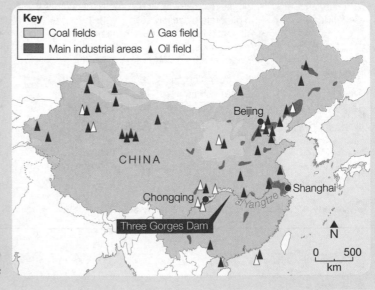

Figure 2 ▶
China's energy resources

- Physical geography determines a country's energy mix.
- Human geography determines how much energy is required and in what forms.
- China is the world's biggest energy producer and consumer, and energy dependent rather than energy secure.
- China's energy mix is varied and projected to change in the future as social, economic and environmental circumstances dictate.

 Over to you

State and learn **five** facts associated with China's energy demands, mix and security.

Student Book
pages 306–7

You need to know:

- about energy supplies in a globalising world
- about Gazprom and the Trans-Siberian Pipeline.

Global energy supplies in a globalising world

Energy is essential in modern-day life, so the geopolitics of energy security is highly significant. Different countries have varying demands, sources and energy mixes, yet compete in a globalising world. Consequently, political friendships and differences, and the power of TNCs are crucial. For example:

- Many of the largest reserves of stock energy resources are concentrated in politically unstable parts of the world (e.g. the Middle East).
- Seven of the world's biggest TNCs (2015) are energy producers, processors and distributors.

Gazprom and the Trans-Siberian Pipeline

Russian TNC Gazprom is the world's largest natural gas supply company, providing all gas for bordering countries, and one-quarter of the EU's (Figure 1). These exports are controlled through pipelines, with the Trans-Siberian, crossing Ukraine, being the most important – it delivers 80% of the gas that Russia exports to Western Europe. It is therefore politically sensitive.

Events such as those outlined in Figure 2, are alarming for many European countries that depend on Russia for their natural gas.

However, the European energy markets are crucial for Gazprom. Because of this, it has secured Europe's energy supplies by the construction of the Nord Stream pipeline which opened in 2012. This runs across the bed of the Baltic Sea bypassing Ukraine and Belarus.

KEY
— Nord Stream pipeline
— Yamal–Europe pipeline
— West and Trans-Siberian pipeline
···· South Stream pipeline (cancelled 2014)

Figure 1
Gazprom's natural gas pipelines

 Figure 2 *Gazprom timeline*

1984 Construction completed, yet considered by some as a threat to the balance of energy trade in Europe.

2004 Democratic replacement of Ukraine's pro-Russian government. Russia quadrupled Ukraine gas prices, and then cut off gas completely when the new government refused to pay! This cut European supplies by 40% in some areas.

2008 With Ukraine seeking to join both **NATO** and the EU, supplies were cut again.

2009–10 Further disputes led Gazprom to cut supplies to Bulgaria, Hungary, Poland and Romania.

2014 Gazprom shut down supplies to Ukraine over non-payment of debts. A day later, an explosion was blamed on depressurisation by some, sabotage by others.

 Sixty second summary

- Political friendships and the significance of TNCs in a globalising world are crucial in ensuring energy security for competing countries, each having different demands, sources and energy mixes.
- Gazprom is the world's largest natural gas supply company, controlling 17% of the Earth's reserves.
- Its gas exports to neighbouring regions, including the EU, are controlled through pipelines – the Trans-Siberian being the most important.

 Over to you

Summarise the key theme of this page in no more than **60** words.

You need to know:

- how Azerbaijan's oil industry impacts on the environment
- about the associated distribution networks.

Student Book
pages 308–9

Baku – the start of the modern-day oil industry

Azerbaijan's oil industry started with the discovery of easily exploitable reserves in the western Caspian Sea region, although transporting the oil to markets was difficult. However, by 1890 Baku was the world's busiest port. Further international investment in pipelines and railway networks followed.

Figure 2 on page 308 of student book shows the infrastructure of the oil/gas industry around the Caspian Sea.

Following the 1917 Russian revolution, decades of Communist mismanagement resulted in overproduction with no incentives to boost efficiency or to invest in better technology. This led to falling productivity and environmental mayhem.

The environmental legacy

The Soviet Union's collapse in 1991 revealed:

- Caspian Sea pollution including runoff from chemically polluted rivers
- air pollution from well fires and natural gas flares
- much of the Absheron Peninsula a wasteland of sulphurous residue, oil ponds and a shoreline blackened by oily tides (Figure 1).

The wildlife and human costs

Prolonged environmental degradation had caused:

- reduced Caspian Sea seal populations and fish stocks
- poor soil, unproductive agriculture, unsafe drinking water
- decreased average life expectancies
- rising infant mortality and higher death rates amongst pregnant women.

Is there hope now?

Independence for the Caspian basin states triggered:

- TNC investment in oil exploitation and distribution infrastructure
- improvements to roads, airports, electricity, communications and services
- a vast array of job opportunities.

But what followed was political instability, ethnic conflicts, disparities between living conditions of locals and expatriate oil workers, and also TNCs breaking their environmental promises. Onshore pollution, wildlife decline and human health deterioration are still apparent today.

 Figure 1 *The environmental impact of drilling on the Absheron Peninsula*

Sixty second summary

- The origins of the modern-day oil industry lie in and around Baku, Azerbaijan, in the Caspian Sea region.
- Decades of Soviet mismanagement resulted in overproduction, underinvestment, inefficiency, falling productivity and environmental degradation.
- Post-Soviet political instability, ethnic conflicts, social inequalities, onshore pollution, wildlife decline and human health deterioration persist today.

Over to you

When revising don't underestimate the power of **PQ2R** – **P**review skimming and anticipation; **Q**uestion to identify the main theme; **R**ead the page carefully (your mind will look for answers); **R**eview to check understanding and test recall.

Student Book
pages 310–13

You need to know:

- about strategies to increase energy supply
- arguments for and against nuclear power
- whether renewable alternatives provide the answer to a sustainable future.

Why is there a need to increase energy supply?

Global energy demands will inevitably increase until 2035 owing to:

- population growth and further economic development
- vehicle numbers anticipated to double
- the need to address **energy poverty** (e.g. 1.2 billion people currently without electricity and 2.8 billion reliant upon biomass).

Increasing oil and gas exploration and exploitation (in the Arctic and South Atlantic or through 'fracking') is controversial, especially as consumption is likely to slow by use of alternative technologies and increased energy efficiency.

However, by 2035:

- most demand is still likely to be met by fossil fuels (especially natural gas)
- nuclear power might double, at most, in EMEs and HDEs
- renewable alternatives will more than double.

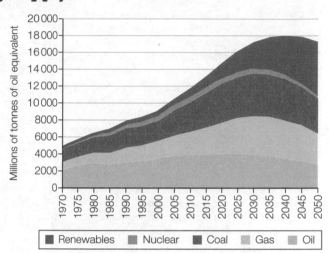

◔ **Figure 1** *Long-term global energy demand predictions*

Nuclear power – keeping an open mind

An extraordinarily powerful and efficient technology.

One of the few low-carbon energy sources already developed.

Safety features in current reactor designs make the chance of an accident extremely unlikely.

The 1986 Chernobyl disaster was caused by a combination of cost-cutting and human error.

Although it might not be an ideal way to meet future energy demands, the dangers of climate change are certainly far worse.

Concerns persist over waste disposal.

Nuclear technology is closely associated with weapons of mass destruction.

The 1986 Chernobyl disaster contaminated huge areas of Ukranian farmland – radioactive rain affected every European country.

Explosions and mass evacuations at the Fukushima Daiichi nuclear power plant followed the 2011 Japanese earthquake and tsunami.

Nuclear power is a very expensive technology.

◔ **Figure 2** *Nuclear power: advantages and disadvantages*

Continued over ▸▸▸

Are renewable alternatives the answer to a sustainable future?

All future projections suggest that alternatives will be, and must be, developed at a much greater rate if future energy needs are to be met sustainably. Each renewable energy source has its own qualities and limitations which make them no less controversial than fossil fuels or nuclear power when considering future energy security (Figure **3**).

Type	Key characteristics and examples	Advantages	Disadvantages
Solar	• Heats water directly or uses photovoltaic cells to convert insolation into electricity e.g. Large developments in California (USA) and Spain e.g Huge potential in the Tropics	• No pollution beyond aesthetics of solar panels • Rapidly falling unit costs of photovoltaic cells • Potential for large developments on agricultural pasture and in deserts	• Expensive on a large scale, despite falling costs • Large areas of photovoltaic cells needed to generate significant electricity • Ineffective in cloud or at night
Wind	• Generates electricity from a rotating turbine, as a single turbine or in wind farms e.g. Onshore; Whitelee, Glasgow e.g. Offshore; London Array, Kent e.g. Significant potential in India and China	• No atmospheric pollution • Abundant exposed sites • UK is one of Europe's windiest countries • Proven technology • Low running costs	• Weather dependent • Noise and visual pollution • Expensive to build, despite falling costs – especially marine wind farms • Bird and marine life affected
Wave	• Oscillating waves force air into a chamber to rotate a turbine e.g. Coastal regions with large fetches, such as Cornwall	• No atmospheric pollution • Largest storm waves during winter high demand	• Wind and weather dependent • High costs of construction and maintenance • Not yet commercially viable • Marine life affected
Tidal	• Uses the incoming and receding water of tidal movement to spin turbines e.g. River Rance, France e.g. UK potential in Severn and Humber estuaries	• No atmospheric pollution • Tides are predictable	• High installation costs • Limited number of suitable tidal ranges • Barrages disturb intertidal ecosystems • Barrages disrupt shipping
Geothermal	• Generates electricity from steam, produced by water being pumped down into volcanically heated sub-surface rocks e.g. Tectonically active regions, such as Iceland, New Zealand and the Philippines	• After initial investment, very low running costs • No pollution	• High installation costs • Limited number of suitable locations

Figure 3 *Solar, wind, wave and geothermal alternatives: advantages and disadvantages*

 Sixty second summary

- Globally, future population growth and economic development dictate an inevitable increase in energy demands.
- Increasing oil and gas exploration and exploitation in the Arctic and South Atlantic or through 'fracking' is controversial.
- Nuclear power is one of the very few low-carbon energy sources already developed – a powerful, efficient, yet expensive and contentious technology.
- Renewable alternatives, such as HEP, solar, wind, wave, tidal and geothermal energy, each have distinct qualities, but also limitations which make them no less controversial than fossil fuels and nuclear power.

 Over to you

This topic is dominated by controversy. Check the advice given on pages 348–50 of the student book, describing how supported judgements and balanced, coherent arguments access the highest levels in mark schemes.

Student Book
pages 314–15

You need to know:

- about strategies to manage energy consumption
- why the German Reichstag is the greenest parliament building in the world
- about decentralising electricity generation by developing combined heat and power systems.

Managing energy consumption

Vast amounts of energy are wasted globally. Electricity generation itself is inefficient – thermal coal-fired stations account for almost 41% of the world's output but they are only 32–42% efficient compared to 85–90% for HEP.

In managing energy consumption it is important to manage electricity generation, by:

- changing the energy mix
- improving energy efficiency and sustainability through nuclear power and renewables.

Governments can encourage change through taxation. For example, the UK Climate Change Levy, introduced in 2001, gives tax incentives to companies adopting energy-efficient practices.

Decentralising energy generation

Decentralising energy generation by developing *combined heat and power* (CHP) systems can be cheap, clean and efficient. It usually involves:

- local generation of electricity
- surplus heat as a by-product which is pumped to consumers as hot water or steam through District Heating (DH) networks of insulated pipes.

Smaller-scale examples CHP and DH systems are found on the Byker Wall estate in Newcastle, which includes three efficient gas-fired boilers and a biomass boiler that burns wood chippings. Residents enjoy remarkably stable heating charges and surplus electricity is sold back to the grid.

 Read about the 'Danish model' on page 315 of the student book.

Figure 1 *In 2018, it was announced that the coal-fired power station at Eggborough, North Yorkshire would close, leaving only six dedicated coal-fired power stations in England and Wales*

The Reichstag – the greenest parliament building in the world

The German parliament building, the Reichstag, was refurbished in the late 1990s to be more energy efficient.

- It uses solar, geothermal, combined heat and power (CHP), biomethane generators, and innovative ventilation.
- More than 80% of its electricity is generated internally; a geothermal installation cools the building in summer and provides heat in the winter
- Special insulation limits heat loss – a 94% cut in carbon emissions.

 Sixty second summary

- Energy is wasted in homes, industries, offices, schools, farms and transport.
- Electricity generation varies in efficiency and is wasteful of primary stock reserves.
- Governments can improve energy management through changing their energy mixes and by adopting creative taxation schemes.
- Innovative architecture and decentralising energy generation by developing combined heat and power(CHP) systems result in remarkable energy savings.

 Over to you

Outline socio-economic and environmental benefits of managing energy consumption.

You need to know:

- about sustainability issues associated with energy production, trade and consumption
- the enhanced greenhouse effect.

Student Book
pages 316–17

What is the enhanced greenhouse effect?

Without the greenhouse effect Earth would be frozen and lifeless! It is climate change resulting from the *enhanced* greenhouse effect that is the problem. Sustainable approaches to energy production, trade and consumption will be key to avoiding the worst consequences of human-enhanced global warming.

 Big idea

The technologies exists to curb greenhouse gas emissions, but is there the political will to develop them further?

Why does it matter?

The Earth is warming abnormally, affecting climates, weather patterns and sea levels, resulting in:

- melting sea ice, continental ice sheets and mountain glaciers
- impacting on landscapes, ecosystems and water security
- rising sea levels threatening low-lying regions and world cities
- more intense rainfall events with localised flooding
- more droughts, heatwaves and wildfires
- stronger hurricanes, cyclones and typhoons
- global environmentally driven migrations (with increased potential for conflict).

⬆ *Amazon rainforest destruction and the enhanced greenhouse effect. Does it matter?*

What can be done?

We already have the technologies, but insufficient political will and investment to curb greenhouse gas emissions.

Except for reducing deforestation and increasing afforestation, all technological 'solutions' involve energy production, trade and consumption. For example:

- increasing public transport availability and car engine efficiency
- converting coal-fired power stations to natural gas, with **carbon capture and storage (CCS)** for all
- increasing the use of biofuels and solar
- doubling nuclear power production
- Increasing efficiency of insulation, lighting and appliances to reduce carbon emissions.

The 1997 Kyoto Protocol

The Kyoto Protocol was the first international agreement committing nations to the reduction of greenhouse gas emissions – and introduced **emissions (carbon) trading** as a practical way forward.

Since then, the 2015 Paris Climate Conference saw 195 countries agreeing the first-ever legally binding global climate deal. It is due to come into force in 2020, setting out a global action plan to limit global warming to 'well below' 2°C.

Refresh your memory of the greenhouse effect on page 316 of the student book.

 Sixty second summary

- The human-caused enhanced greenhouse effect is arguably the greatest environmental challenge of our time.
- Abnormal global warming is affecting climates, weather patterns and sea levels with potentially alarming consequences.
- Reducing greenhouse gas emissions needs sufficient political will and financial investment.
- The 2015 Paris Climate Conference agreed to limit global warming to 'well below' 2°C.

Over to you

Don't assume you already know the ins and outs of the greenhouse effect. Use the Link and check!

Student Book
pages 318–19

You need to know:

- about energy conservation, acid rain and nuclear waste.

Energy conservation

Energy conservation is a key issue in reducing greenhouse gas emissions, addressing energy poverty, ensuring sustainability and saving money.

More than a quarter of UK CO_2 emissions come from domestic energy use, so greater efficiency is encouraged by increasingly demanding UK building regulations, such as:

- improving thermal efficiency of windows (double/triple-glazing), walls and roofing
- government grants for fitting cavity wall and loft insulation in existing buildings
- clearly labelled energy-efficient domestic appliances and light bulbs
- 'smart' electricity meters using real-time displays of consumption
- adopting solar water heating and electricity-generating photovoltaic panels.

Acid rain

Acid rain caused by the burning of fossil fuels was common in Europe in the 1980s and 90s. It is still of concern in China, India, Thailand and South Korea.

Nuclear waste

Nuclear power is a 'clean', sustainable, albeit expensive solution to energy emissions and security issues, particularly in EMEs and HDEs (see 5.12, 5.16).

But nuclear power stations produce radioactive waste fuel rods, which have to be carefully managed. This waste has a long half-life and so safe transport and storage is essential.

The THORP reprocessing plant at Sellafield in Cumbria processes used fuel rods from all over the world and stores the remaining waste in steel-clad or concrete and lead-lined glass containers.

Waste stockpiles at Sellafield cannot increase indefinitely, so the need for geologically stable, accessible, safe and secure long-term storage sites is pressing.

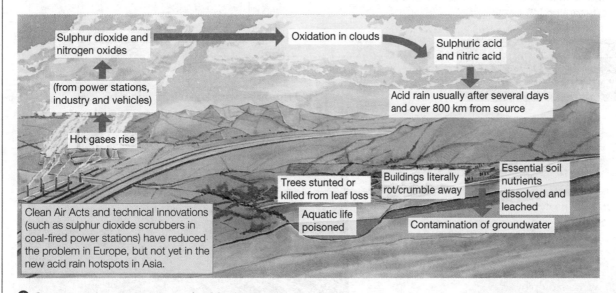

Sulphur dioxide and nitrogen oxides

(from power stations, industry and vehicles)

Hot gases rise

Oxidation in clouds

Sulphuric acid and nitric acid

Acid rain usually after several days and over 800 km from source

Trees stunted or killed from leaf loss

Buildings literally rot/crumble away

Essential soil nutrients dissolved and leached

Aquatic life poisoned

Contamination of groundwater

Clean Air Acts and technical innovations (such as sulphur dioxide scrubbers in coal-fired power stations) have reduced the problem in Europe, but not yet in the new acid rain hotspots in Asia.

🔺 *Causes and consequences of acid rain*

Sixty second summary

- Energy conservation is a key issue in reducing greenhouse gas emissions, addressing energy poverty, ensuring sustainability and saving money.
- UK building regulations encourage energy efficiency promoting truly sustainable, low-carbon homes.
- Acid rain is **trans-boundary pollution** with solutions that have not yet been adopted in Asia's pollution hotspots.
- The need for geologically stable, accessible, safe and secure long-term storage sites for high-level nuclear waste is pressing.

Figure 1 on page 318 of the student book gives an example of a sustainable, low-carbon home.

Over to you

List actions you could take, or have already taken at home to save energy.

Student Book
pages 320–5

You need to know:

- sources, demand, distribution and uses of copper
- the physical geography and geological conditions associated with copper ore occurrence
- environmental impacts and sustainability of extraction, processing and distribution.

Copper's importance

Copper is one of the world's most important and widely used metals (Figure **1**). It is durable, malleable, oxidises (corrodes) very slowly and totally recyclable, with nearly 90% of the available scrap currently recycled.

Sources, distribution and trade

Both recycling and mining is needed to meet an annual consumption that exceeded 21 million tonnes in 2015. Ore deposits are highly concentrated in the veins and cavities of basic igneous rocks distributed globally (albeit particularly in the Americas). In almost every location, the metal content is very low (< 0.5%) and so only open-pit mining and primary processing on site is economic.

Latin America (especially Chile and Peru) accounts for almost half of global copper production (Figure **2**). But more than 60% is consumed in Asia, necessitating a global trade in concentrates, anodes, cathodes, ingots, semi-finished products and scrap for recycling. China consumes 42% of global copper and is now the world's leading manufacturer, having overtaken the USA in 2015.

See Figure 4 on page 321 of the student book for a map of global copper distribution.

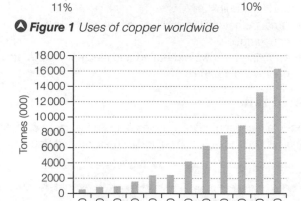

Figure 1 *Uses of copper worldwide*

Electronic goods 38%
Construction 31%
Household appliances 10%
Industrial equipment 10%
Transport 11%
Cu

Figure 2 *World copper mining production*

The Kennecott Bingham Canyon Mine, Utah, USA

The Kennecott Bingham Canyon Mine, 30 miles south-west of Salt Lake City, is the largest copper mine in the USA. Primarily copper, but precious metals too, are extracted from the world's deepest man-made quarry – 4.5 km across and 1.2 km deep (Figure **3**).

Scale is everything in this facility:

- 2000 workers employed
- 450 000 tonnes of rock extracted daily
- 500 miles of internal roads
- US$1.8 billion worth of metals produced annually
- more than 19 million tonnes of copper has been produced – more than any other mine in history.

The UK's Rio Tinto Group bought Kennecott in 1989 and has invested nearly US$2.5 billion in modernising the mine and addressing environmental problems. It has also opened a visitor centre to educate the public about mining practices, sustainable development and the importance of mining in modern life.

Figure 3 *Aerial view of the Kennecott Bingham Canyon Mine*

Environmental impacts of copper mining

Severe environmental impacts include:

- disruption of the landscape
- removal of vegetation and topsoil
- contamination of air with dust and toxic substances
- groundwater contamination by toxic compounds (including acids) in mine *tailings*.

Processing also has environmental impacts. The ore is low in metal content so there are large amounts of both *overburden* (waste material lying over the ore) and *tailings* (materials left over after extraction).

This waste can cause environmental problems if simply dumped on spoil heaps or back-filled into the pit. If the tailings are not covered and stabilised, dust and water leaching through the waste can carry toxic materials into the environment.

Consequently, dams, embankments and other types of surface impoundments (Tailings Management Facilities) are commonly used for storage. Air pollution from both smelting gases and dust also require management (Figure **4**).

Impact	Management
Dust from blasting and also trucks carrying ore, tailings and stockpiles	Mine sites and roads sprayed with water (recycled whenever possible)
Overburden dumped in spoil heaps	Spoil heaps drained and limited in size before recontouring and landscaping with topsoil and vegetation
Tailings impounded in dammed tailings ponds	Tailings ponds eventually covered with clay, topsoil and revegetated
Large pits created from open-pit mining	Exhausted pits filled with groundwater (for recreational lakes), rock waste or MSW landfill (see 3.20)
Sulphur dioxide gas produced during smelting	Most collected and used to make sulphuric acid for use in fertiliser

⬆ **Figure 4** *Managing the impacts of copper mining*

Sustainability in the copper industry

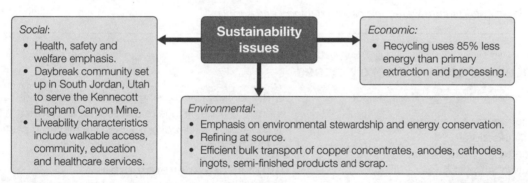

Social:
- Health, safety and welfare emphasis.
- Daybreak community set up in South Jordan, Utah to serve the Kennecott Bingham Canyon Mine.
- Liveability characteristics include walkable access, community, education and healthcare services.

Sustainability issues

Economic:
- Recycling uses 85% less energy than primary extraction and processing.

Environmental:
- Emphasis on environmental stewardship and energy conservation.
- Refining at source.
- Efficient bulk transport of copper concentrates, anodes, cathodes, ingots, semi-finished products and scrap.

⬆ **Figure 5** *Sustainability issues associated with copper extraction, trade and processing*

 Sixty second summary

- Copper is a key raw material in electricity transmission, electronics, telecommunications, construction, transport, household appliances and industrial equipment.
- Recycled copper accounts for one-third of all copper used.
- Latin America accounts for almost half of global copper production, but China consumes 42% of global copper.
- Rio Tinto Group's Kennecott Bingham Canyon Mine has produced more copper than any other mine in history.
- Copper mining can have severe environmental impacts, but careful management, environmental restoration and stewardship can demonstrate sustainability.

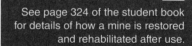 See page 324 of the student book for details of how a mine is restored and rehabilitated after use.

 Over to you

Carefully consider the paradox of sustainable copper production. Outline the evidence you might use in support of the argument that it is, in fact, sustainable.

Student Book
pages 326–7

You need to know about:
- factors influencing future water and energy demands and supplies
- water and energy futures in India.

Supply and demand – the future

How nations respond to future challenges of supply and demand may reflect one of three scenarios:

1 Business as usual – no crisis so no changes required.

2 Technology, economics and **privatisation** – the **market economy** will solve any problems using technology.

3 Values and lifestyles – shift in attitudes, needs education, international cooperation and behavioural changes.

Factors influencing water/energy futures	Effect on water supply and demand	Effect on energy supply and demand
Technological	• GM crops – need less water • Desalination plants increase supply	• New energy reserves become exploitable using new technologies (e.g. fracking)
Economic	• Reduction of water footprint – as a result of rising costs of water supply • Expensive engineering projects	• Previously uneconomic resources become viable (e.g. Arctic oil) • Finite resources become more costly as supplies run out
Environmental	• Improvements in efficiency and waste reduction (e.g. rainwater harvesting)	• Fragile environments (e.g. Alaska, Antarctica) under threat as existing reserves exhausted • Climate change encourages reduction in use of fossil fuels
Political	• Fewer dams will be built if opinions influence decision-making • Less food miles – to reduce water footprint	• NIMBYism – continues to influence policy-making (e.g. strength of opinion towards nuclear energy and wind power)

🔺 **Figure 1** *Factors affecting future water and energy demand and supplies*

India's future water and energy crises

Factors in India's water and energy crises include:

- rapid population growth
- high rates of urbanisation
- large population size
- effects of climate change
- huge and growing economy
- increased consumerism.

Demand for water is increasing across India from:

- industry – a result of continuing economic growth
- urban consumers – supplying clean water to areas of high population density is challenging
- growing middle class – aspire to more 'western' clothing and diets, production of which is more water intensive.

Whilst demand is increasing, there are uncertainties over supply – climate change (e.g. failed monsoons), leads to regional uncertainties of supply. Environmental impacts also include widespread pollution of surface water.

India's future energy crisis

The part privatisation of India's energy sector has not fully resolved issues of:

- under investment (e.g. in HEP megaprojects or nuclear power)
- increased reliance on energy imports (e.g. one-third of gas is imported)
- potential geopolitical conflict (e.g. gas pipelines have not been built with neighbouring Pakistan and Iran).

Sixty second summary

- Future global water and energy supplies will struggle to meet demand.
- The factors influencing both water and energy futures are technological, economic, environmental and political.
- India is facing water and energy crises caused by its rapidly growing population and economy, increased consumerism, urbanisation and also by the effects of climate change.
- India's threatening water and energy crisis is resulting in significant economic and environmental challenges.

Over to you

Create a mind map that summarises factors influencing future water and energy demands and supplies in India. Give reasons why you are optimistic or pessimistic over water and energy futures in India and globally.

Student Book
pages 328–31

Case Study

You need to know:

- about California's water crisis at regional and local scales.

Crisis ... what crisis?

California has a population of 39 million. If it was a country, it would be the seventh largest economy in the world. California has a demand for water that exceeds natural supplies. This demand is increasingly more difficult to meet:

- there is a **spatial imbalance** – three-quarters of California's growing population lives to the south of Sacramento, yet three-quarters of the precipitation falls to the north (Figure **1**)
- 50% of precipitation falls between November and March
- 65% of precipitation is already lost to evapotranspiration – with climate change increasing the variability and unreliability of this rainfall even more
- droughts are frequent – 2015 was the fourth consecutive year of drought.

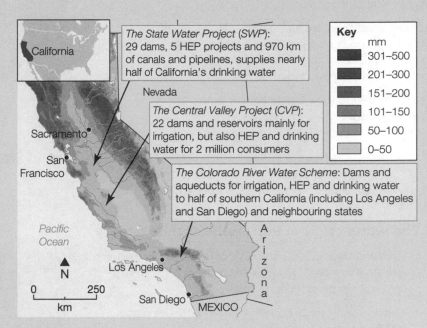

The State Water Project (SWP): 29 dams, 5 HEP projects and 970 km of canals and pipelines, supplies nearly half of California's drinking water

The Central Valley Project (CVP): 22 dams and reservoirs mainly for irrigation, but also HEP and drinking water for 2 million consumers

The Colorado River Water Scheme: Dams and aqueducts for irrigation, HEP and drinking water to half of southern California (including Los Angeles and San Diego) and neighbouring states

Key
mm
- 301–500
- 201–300
- 151–200
- 101–150
- 50–100
- 0–50

Figure 1 *California's annual precipitation and water supply systems*

Water security – challenges, conflicts and controversies

The majority of California's supply normally comes from ten drainage basins. Groundwater (aquifers) accounts for 30% and a small proportion from desalinated sea water. Groundwater demands rise by up to 60% in drought years and aquifer recharge has not kept pace with extraction.

Hence the importance of the world's largest interconnected, though controversial, water network. This local, state and federal funded network distributes water across the state and into neighbouring Nevada and Arizona (Figures **1** and **2**).

Ensuring water security in California raises significant challenges of conflicting demands, conservation issues and questions of 'ownership'. For example:

- 80% is used for farming – mostly using efficient sprinklers and drip, but some flood irrigation persists.
- Environmentalists are concerned that wetlands have already been drained, natural habitats altered and river fish stocks depleted.
- Water rights disputes – with bordering Mexico, with native American tribes and within the state (north–south 'ownership' and supply concerns).

In 2015, mandatory restrictions to cut demand by 25% were introduced.

- Higher rates and fees discouraged water waste.
- State-sponsored aid for homeowners who install more efficient garden watering systems.
- Conservation promoted through the use of greywater, drought-tolerant landscaping, and so on.

Figure 2 *The Hoover Dam, part of The Colorado River Water Scheme*

Continued over >>>

Case Study

Good enough to drink?

Throughout Los Angeles County groundwater has been contaminated for decades by industrial, agricultural and domestic waste water seepage (Figure **3**). The most polluted wells have had to be closed. The Colorado River Water Scheme has, until recently, met demand.

But now water agencies are having to make use of groundwater sources again. Most communities in southern LA County have to rely, at least partially, on contaminated groundwater supplies.

Treatment is expensive, but a cheaper option than importing water.

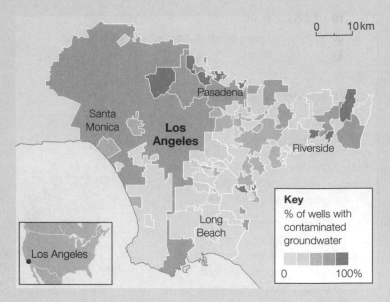

Figure 3 ▷
Percentage of water wells with contaminated groundwater (prior to purification), Los Angeles County

Key
% of wells with contaminated groundwater

0   100%

The water crisis in Porterville, Tulare County

In 2015 and 2016, 55 000 residents of Porterville were reliant on private water wells. Many of these wells dried up, so these people were dependent on:

- community roadside containers for toilet flushing and washing clothes (Figure **4**)
- a county-wide Bottled Drinking Water Programme.

Tulare County lies in the heart of California's rich agricultural belt. Labour-intensive hand-picked fruits and vegetables have traditionally employed many farm workers, but the prolonged drought has reduced agricultural water allocation.

Farmers, in turn, have left land fallow or chosen to grow smaller quantities of less water-intensive, higher value machine-harvested crops, such as almonds and pistachios.

The resulting 19% reduction in farm workers is causing a negative socio-economic multiplier effect for whole communities:

- one in four families in Tulare County now live below the federal poverty level – reliant on food stamps and aid
- families are being forced to relocate to find work
- services such as shops are losing business; schools are losing staff.

Even when and if the rains return, the shift to more mechanised forms of farming may mean recent changes to agriculture become permanent.

⬆ **Figure 4** *Community water tank in East Porterville, Tulare County*

🕐 **Sixty second summary**

- California's greatest geographical threat is water supply; ensuring water security and long-term sustainability of supply is a challenge.
- The demand for water already exceeds supply and there is a spatial imbalance between the highly populated, drier south and the less populated, wetter north.
- California's controversial water supply and transfer system is the world's largest and most productive scheme of its kind.
- Industrial, agricultural and domestic contamination of groundwater supplies in Los Angeles County and acute water supply problems in Tulare County result in significant socio-economic costs.

✏ **Over to you**

Sequence the negative socio-economic multiplier effect provoked by prolonged drought in Tulare County.

Student Book
pages 332–3

Case Study

You need to know:

- about the effects of the physical environment on oil exploration and exploitation in Alaska
- what controversies surround the industry's future.

Alaska – physical geography

The vast US state of Alaska is:

- tectonically active – bordering the 'Pacific Ring of Fire'
- glaciated in the north and periglaciated further south; it is warming significantly owing to climate change
- a climatically and ecologically diverse wilderness, with protected tundra reserves of fragile flora and fauna, and wildlife refuges.

Oil exploitation in Alaska

Alaska is a *single-product economy* with oil exploitation providing one-third of all jobs. It funds up to 90% of its operating budget.

The oil boom began in the 1970s with the development of the 25 billion barrel, ice-bound Prudhoe Bay oilfield and revolutionary Trans-Alaska pipeline, which:

- crosses two mountain ranges and 800 rivers
- is raised and insulated to prevent the hot oil melting the permafrost, while allowing caribou to migrate underneath
- slides during earthquakes, but with automatic shut-off systems if there is a leak.

However, since peaking in 1988, oil production has declined, prompting oil security and human welfare arguments about further onshore development or pushing technological extremes to increase offshore drilling (Figure **2**).

 **Big idea**

Alaska's socio-economic geography is dominated by the exploitation of its enormous oil reserves.

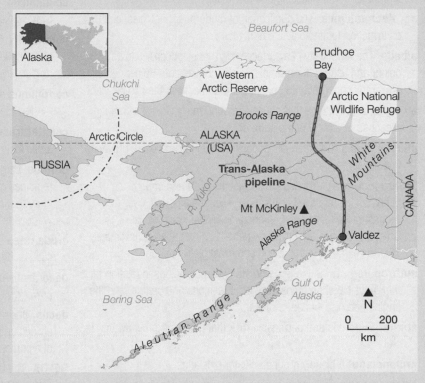

⬆ *Figure 1* Alaska has a diverse wilderness and protected environments. The Trans-Alaska pipeline connects the northern oilfields with the port of Valdez in the south.

For (mainly socio-economic)	Against (mainly environmental)
• The Arctic National Wildlife Refuge has estimated reserves of over 10 billion barrels	• Native Americans depend on wildlife for their food, clothing and shelter
Development of Chukchi Sea reserves would: • secure and create jobs • provide much-needed oil for the Trans-Alaska pipeline	• Arctic waters could be spoiled • Directional drilling might reduce, but not eliminate the impact of road, plant, accommodation and other infrastructure
Exploitation would further US energy security; lower oil prices for consumers; increase tax revenues; reduce the US trade deficit	• An untouched wilderness of fragile tundra soils and vegetation, on already deteriorating permafrost, might never recover

⬆ *Figure 2* Arguments for and against further development of Alaskan oil reserves

 Sixty second summary

- Alaska has contrasting physical geography and a rapidly warming climate.
- 90% of Alaska's operating budget and one-third of all jobs are directly dependent upon the oil industry.
- Oil security and human welfare are inextricably linked, but the industry's future is uncertain and controversial.
- The Arctic National Wildlife Refuge is a virtually untouched wilderness with significant oil reserves.
- Plans to exploit reserves in the Chukchi Sea were abandoned in 2015.

 Over to you

What economic reasons might have influenced abandonment of plans to exploit Chukchi Sea oil reserves in 2015?

Glossary

actors Players on the world stage – the governments and other global institutions, both public and private, that participate in global governance, exercise power, make decisions, solve problems and improve lives

agglomeration economy The economies of scale derived by firms clustering together and sharing ancillary services and public utilities; they also include industrial linkages, the development of a specialised labour force, bulk buying and shared marketing

agribusinesses Large farms, plantations or estates owned and managed by major companies that organise the purchase of all inputs, labour, processing and marketing

agrotechnologies The application of modern technologies, such as irrigation technologies to agriculture

albedo The reflectivity of a surface which varies according to its colour and texture; for example, fresh snow reflects 85 per cent of solar radiation, grass 25 per cent, but tarmac less than 10 per cent

allergen A substance that causes an abnormally vigorous immune response or allergy in the sufferer

alluvial plains Sedimentary deposits ranging from coarse gravel to fine silt spread across floodplains

Anthropocene A term used to convey the scale and importance of human influence in Earth's most recent time period, given widespread evidence of anthropogenic global warming etc. Scientists suggest that this new period of time began approximately 10 000 years ago with the end of the last glacial period.

anthropologists Scientists who study human beings in relation to their physical characteristics, social relations and culture, and the origins and distribution of races

aquifer A porous and permeable rock that acts as a groundwater store

archipelago A closely grouped cluster of islands

artesian basin Low-lying region where groundwater is confined under hydrostatic pressure from surrounding layers of rock; often found where an aquifer lies trapped in a syncline

asylum seeker A person who is seeking international protection but whose claim for refugee status has yet to be determined (see refugee)

bandwidth throttling The deliberate slowing of internet service by an internet service provider (ISP)

big data Contemporary term used to describe extremely large datasets from which we can learn a great deal with effective analysis

bilateral Between two parties

bioprospecting The process of discovery and commercialisation of new products based on biological resources

carbon capture and storage (CCS) Technology designed to replace the pumping of industrial and power-station fossil fuel CO_2 into the atmosphere

carrying capacity The idea of a population ceiling beyond which an environment cannot support people at a high standard of living (or at a subsistence level), for a sustained period of time without environmental degradation

channelisation Straightening and lining of a river channel to improve its rate of flow and navigability

chronic A term used to describe an illness that lasts for a long time or is constantly recurring

COBRA Acronymn for Cabinet Office Briefing Room A; the government emergency response committee, including government ministers, police and intelligence officers

code Title or abbreviation of a title used by a researcher to label and sort sections of text, according to theme to aid analysis; different codes and sub-codes may be applied to individual paragraphs or lines of text within a qualitative data source (e.g. an interview transcript)

colonialism The policy or practice of a power in extending control over weaker nations or peoples

community-focused regeneration Regeneration focusing on the social needs of communities; for example, affordable housing to rent or buy, shops and schools

continuous data Data that can take any value (within a given range), such as life expectancy

counterurbanisation Population movement from large urban areas to smaller urban settlements and rural areas

cratons Large, ancient sections of the Earth's crust that have remained relatively stable for considerable periods of geological time; they are associated with the 'drifted' fragments of Alfred Wegener's 'supercontinent' Pangaea

crude rates A measure of the basic statistics of any population, such as birth or death rates per 1000

decentralisation Process of redistributing people, functions or power away from the centre to the periphery

decile One of ten equal subsections that a population may be divided up into, according to the distribution of the ranked values of a particular variable

DEFRA The UK government's Department for Environment, Food and Rural Affairs

deindustrialised Long-term decline of industry leading to significant social and economic changes

demographers People who study population

demographic dividend A falling birth rate results in a smaller population of young, dependent ages and relatively more people within the economically active adult age groups; this improves the ratio of productive workers to child dependents which can encourage economic growth

deprivation A meaningful measure of poverty, defined in terms of people's lack of access to social and economic necessities. The Townsend Index of material deprivation (1988) incorporated four variables: unemployment; non-car ownership; non-home ownership; and household overcrowding.

dew point The critical temperature at which air, on cooling, becomes saturated with water vapour; as temperature falls further the vapour condenses into cloud droplets; consequently, the dew point is defined by the cloud base

dialect A particular form of a language which is peculiar to a place, region or social group

disability-adjusted life year (DALY) A measure of the overall burden of a disease that combines mortality (early death) and morbidity (ill-health, disability) in a single measure

Glossary

discrete data Data that can only take certain values, such as number of babies born to a woman

diseases of affluence Degenerative diseases such as cancer, heart disease and dementia most associated with lifestyles in more economically developed countries

drosscape A landscape of industrial dereliction; rapidly urbanised regions that become the waste products of historic, economic and industrial processes

Dutch disease The negative consequences as a result of large increases in a country's income; it is usually associated with the discovery of natural resources, particularly oil reserves, but can also result from any large increase in foreign currency such as FDI

ecological footprint A measure of the demands we humans place on ecosystems which support us; the amount of biologically productive land that is used to produce the resources we consume and to absorb the waste we generate

economic migrant A person who has left their own country to seek employment in another country in order to improve their living conditions

edge cities Modern suburban areas that act as an alternative central business district, including shops, offices and entertainment; they are characteristic of low density suburbanisation such as in the USA

emancipated Free from domestic or cultural restraints and so allowed to pursue independent lifestyles and careers

emissions (carbon) trading Effectively trading a permit to pollute. These permits are distributed to polluting organisations that may only emit as much carbon as it has allowances for – hence having to buy the right to pollute from more efficient businesses if it cannot become more environmentally friendly.

empowerment To give power or authority to someone; female empowerment involves the fuller participation of women in a nation's economy and society

endemic Disease that is always present in a population

endogenous factor A key aspect of a place's local geography (physical or human) that helps to shape its unique character, for example, geology

energy dependent The higher the proportion of energy imported, the more energy dependent the country is on others. In such cases, a diversification of both energy sources and suppliers is crucial.

energy poverty Having less energy than is required to meet demand

energy security The uninterrupted availability of energy sources at an affordable price. For example, Russia is very energy secure because of its huge energy surplus. The UK is energy insecure because of its energy deficit and has to import much of its supplies.

epidemic Widespread occurrence of an infectious disease in a community at a particular time

epidemiological transition The theory that populations, typically, shift from being defined by high rates of infant mortality (and low life expectancy) to a state in which average life expectancy is much higher (50+ years)

exogenous factor A relationship with another place/s that help to shape the unique character of a place, for example, membership of the European Union; such relationships can be seen in the movement or flow of people, resources, money and ideas across space

experienced place A place experienced directly by the individual

exponential growth Increasingly more rapid growth at a constant rate

exurbs Residential areas that are planned and built beyond the suburbs to create large extended suburbs, typically associated with urban sprawl

Fairtrade A value-based organisation and trademark that aims to tackle injustices of the globalised economy. Fairtrade aims to pay farmers a guaranteed minimum price, offer fair terms of trade and make payment of an additional development premium for reinvestment.

far (distant) place Somewhere that an individual or society perceives as being physically distant, generally inaccessible. Beyond actual spatial distance, such a perception may be shaped by networks of infrastructure (transport, communication) or access to them. Moreover, such a place may be viewed by some people as being different, even alien or exotic.

flexibility of production A method of production that is sufficiently flexible to be able to respond to both planned and unplanned changes, such as strikes or natural disaster

forces of change The individuals, community groups, companies, governments, national and international institutions that influence the place-making process

Fordism System of mass production of goods that involved assembly lines and 'living wages' for employees, pioneered by Henry Ford In the manufacture of cars in the early twentieth century

foreign direct investment (FDI) An investment made by a company, usually a TNC, based in one country into a company based in another country; the investment is usually made to acquire control or to have significant influence over the foreign company

fracking More correctly known as hydraulic fracturing. Oil- and gas-bearing shale is drilled and fractured by high-pressure injection of water, sand and (toxic) chemicals. Cracks are created in the shale through which the oil or gas will flow more freely.

fuel poverty The condition of being unable to afford to heat your home sufficiently, given your level of income and all other outgoing costs, such as food, rent, transport, etc.

gatekeeper Term used by social scientists to describe individuals within a community or institution who can grant a researcher access to people or sources of data by virtue of their economic role or social standing

generic medication A drug that is comparable to a well-known, branded drug in form, strength and performance but that is produced by a different manufacturer, and is cheaper for the patient or health service to purchase

genius loci A term used by planners to describe the key characteristics of a place, with which any new developments must concur

gentrification The improvement of urban areas by individual property owners, which usually leads to increased commercial activity in local retail areas

geopolitics The study of the ways in which political decisions and processes affect the use of space and resources; it is the relationship between geography, economics and politics

geospatial data A type of data that has a spatial or geographic component, meaning it can be mapped; pieces of digital data may have explicit geographic positioning (e.g. latitude and longitude) information linked to them, such as georeferenced satellite images

ghetto An area of a city in which large numbers of people of a particular minority ethnic group live

global product A product that is marketed and branded throughout the world. Many TNCs produce global products, for example, Coca-Cola, Nike and (Jaguar) Land Rover.

heritage tourism Travel to experience places, artefacts and activities that represent the stories and people of the past; the degree to which such an experience may be 'authentic' is a matter of debate

horizontal integration Improving links between different firms in the same stage of production

identity Who a person is, both in terms of how others view them, and how they see themselves. A person's identity is shaped, in part, by where they live and/or their place of birth (their homeland).

index case First identified case of a disease

indigenous Originating in a particular region or environment

infrastructure Basic facilities and installations that allow a city/country to function

insider Someone who feels safe, secure and 'at home' in a place; they understand the social norms of the society and feel included. They can play an active social and economic role in society.

knowledge economy An economy based on creating, evaluating and trading knowledge and high level skills

leachates Toxic waste water containing arsenic, lead, solvents and other contaminants leached from illegal dumps and landfill

leaching The process by which heavy rainfall infiltrates through a soil, removing humus and nutrients in solution

liveability The combination of factors that determine a community's quality of life; they include the built and natural environments, its accessibility, economic prosperity, social stability, educational opportunities and sustainability, and culture, entertainment and recreation

locale A setting where everyday life activities take place, for example an office, park or cruise ship; people behave in a certain way in a locale, according to social norms or rules

location A physical position that can be plotted on a map

malnutrition Inadequately balanced diet whether through undernutrition or overnutrition (obesity)

marine protected areas (MPAs) Marine areas where certain activities are limited or prohibited in order to meet specific conservation, habitat protection, or fisheries management objectives

marine reserves Marine reserves are fully protected areas that are off-limits to all extractive uses, including fishing. They provide the highest level of protection to all elements of the ocean ecosystem.

market economy An economy in which economic decisions, such as those regarding investment and production, are based solely on supply and demand with little government involvement

media place A place experienced only through the media

megaprojects Very large investment projects that typically cost more than (US)$1billion

metropolitan Spatial area that is greater than the limits of the city it relates to; it includes both the densely-populated urban core and its surrounding suburbs (plus, in some cases, other smaller urban areas) that are bound to the city by employment, commerce and/or infrastructure

microclimate Climate within a relatively small area that is distinctively different from the climate of the surroundings; for example, the climate of an urban area contrasting markedly with that of the surrounding countryside

Millennium Development Goals Eight international development goals that were established following the Millennium Summit of the United Nations in 2000; each goal had a target to meet by 2015. Replaced by the Sustainable Development Goals in the same year.

modes Particular forms or means of something, for example, transport

monopoly When a single company or group owns all or nearly all of the market for a given type of product or service – there is little choice and little competition

morbidity The incidence or prevalence rate of a disease or all diseases within a population

mortality Being subject to death, being mortal; the most common indicator of mortality within a population is the death rate

myth Socially-constructed versions of reality that may, as a result of its long history or because it is widely-held, be thought of as common sense

natural population change The pattern of change in population over time that does not take into account the impact of migration

near place Somewhere that an individual/society perceives as being physically close either by virtue of being easily accessible and/or spatially close. Such a place may be viewed as being similar, perhaps, even inextricably linked to the place where an individual/society is located.

neo-Malthusianism Views or attitudes that are in common with Thomas Malthus who believed that there are environmental limits to population growth

NIMBYism 'Not in My Back Yard'. An attitude shared by those who do not want a development, such as a wind farm, in their near locality

non-utilitarian Not practical, such as a beautiful landscape

North Atlantic Treaty Organisation (NATO) Based upon the North Atlantic Treaty signed in April 1949. It is an intergovernmental military alliance whereby its member states agree to mutual defense in response to an attack by any external party.

obese For adults of both sexes, aged 18 and over, obesity is a body mass index (BMI) greater than or equal to 30 (overweight is a BMI between 25 and 29.9)

Office for National Statistics (ONS) largest producer of official statistics in the UK

optimum population A theoretical concept, and ideal; the number of people that can make the best use of all available resources within a country or region, ensuring that everyone has an adequate standard of living

ores Rocks where the mineral content (usually metal) is of sufficient economic value to justify exploitation

Organisation of Petroleum Exporting Countries (OPEC) An organisation or cartel that follows a common approach to the sale of oil

Orientalism Edward Said's theory that, historically, Europeans viewed people in the Orient (a region that included the Middle East, North Africa and Asia), as being not only exotic, but decadent and corrupt. Such a view of the Orient was used, at the time, to justify the actions of imperial powers.

other Someone or something that is different, alien or exotic. A person living in a distant place (or key characteristics of their way of life) may be defined as 'other' by individuals or a society, as a result of the perceived contrast between 'them' and 'us'.

out of place A feeling of not being 'normal', not fitting in, in the context of a particular society or locale

outsider Someone who feels homesick, alienated or excluded from society in a specific place; they may not be able to take an active role, for example, in work or study as a result of socially-constructed barriers

overpopulation Too many people for the resources or technology available in a given area, or country, to support at an adequate standard of living

pandemic An epidemic occurring worldwide, or over a very wide area, crossing international boundaries and usually affecting a large number of people

particulates Tiny particles, such as dust or soot, given off when fossil fuels such as coal or oil are burned

place More than its physical location, a place is a space given meaning(s) by people

place identity The real or perceived characteristics of place that people articulate; the 'social narrative' or the stories people tell about a place

place study An investigation of a place and its developing character, involving the collection, analysis and interpretation of both quantitative and qualitative data, including representations of it in the media

placelessness The idea that a particular landscape 'could be anywhere' because it lacks unique features. Some UK high streets have been criticised for being dominated by identically-branded chain stores.

planning blight The reduction of economic activity and/or property values in a particular place resulting from expected or potential future development or restrictions on development

postmodernism A philosophical movement that applies a particular viewpoint to the world in which we live

primary health care Health care that is provided at the point of contact to all in the community

primary products Goods made up of a natural raw material and that have not been through any manufacturing process (such as oil, timber or fish)

privatisation Transfer of ownership from the government to private companies and businesses

property-led regeneration Regeneration led by property developers and financiers with little input from government except the creation of opportunities or tax breaks

provenance The context in which a source or text is produced, which may give clues to its purpose (and its reliability)

qualitative data Data that can only be organised into descriptive categories that are not numerical; such data may include oral sources such as interviews, reminiscences and songs, and visual media include artistic representations

QUANGO An acronym for quasi-autonomous local government organisation; an administrative body funded, but not run, by the government

quantitative data Numerical data to which different techniques of statistical analysis may be applied to test a hypothesis

rebranding A process to rename and reimage a location to boost footfall, investment and the wider social and economic prospects of a place

refugee A person who has been granted leave to stay in a foreign country, having been forced to leave the country of his or her nationality, 'owing to well-founded fear of being persecuted for reasons of race, religion, nationality, membership of a particular social group or political opinion' (UNHCR)

regeneration The process of urban or rural improvement, which may be economic, social or environmental in nature

reliable A social scientist must try to evaluate how accurate or useful a particular source is. This can be done by comparing it to other sources and also finding out about the author or artist, their wider views and those of their patrons (its provenance).

remote sensing The scientific collection of (mass) data from objects or areas without being in physical contact with them; the data is gathered by electronic scanning devices carried in high-flying aircraft or satellites

residential differentiation Demarcation of socio-economic, cultural and ethnic groups within specific housing areas

retrofit To adapt something or add to it, after it has been installed for use, for example, the addition of double glazing to a house once built and single-glazed

reverse osmosis Involves pushing salt or brackish water through a porous membrane that filters out salts and other impurities to produce freshwater

royalties Proportion of profits paid to whoever grants a mining lease

Rust Belt A region that stretches across the north-east of the USA, the Great Lakes and the Midwest states that has suffered from economic decline of key industries; it was once called the Steel Belt

sedentary Too much sitting and insufficient physical exercise

sense of place (place-meaning) An individual's subjective and emotional attachment to a place, its place-meaning

Glossary

single-product economy A country which relies on one, or a very small number, of products (usually raw materials) for its export earnings

socially constructed Social processes produce and reproduce the social and economic relations between different groups of people in society, in different locations. One aspect of this is place-meaning. The dominant place-meaning of a location may benefit dominant classes and the status quo.

space The three-dimensional surface of the Earth; further defined by human geographers as a container in which objects are located and human behaviour is played out

spatial imbalance An unevenness in any geographical distribution, for example, precipitation

sports-led regeneration Regeneration which focuses on the benefits brought by sports events

spreading of locational risks Spreading of risk by investing in different geographical locations – 'not putting all the eggs in one basket'

stakeholders Individuals, groups or organisations that are affected by an issue

stewardship The responsible use and protection of the natural environment through conservation and sustainable practices

suburbanisation The outward growth of people, services and employment towards the edges of an urban area

syncline Where rocks have been folded downwards into a basin shape

teleworking A work arrangement that allows an employed person to work from home; this is made possible by the use of technologies such as the internet, satellite technologies and mobile phones

terms of trade The value of a country's exports relative to that of its imports

text A source for analysis; texts used by cultural geographers include works of art and films, as well as written sources like novels, poetry and travelogues

thermal distillation Uses heat to create a vapour from salt or brackish water which is then converted into freshwater

tourist gaze What a visitor sees or experiences of a place of interest (e.g. a historic site). To some extent, it is organised or edited by professionals in the tourist industry.

trade bloc Where a set of countries trade freely with each other with few, if any, barriers such as tariffs. Countries outside this area that wish to trade anywhere within the trade bloc have to pay an agreed tariff.

trade protectionism The use of barriers such as import duties (tariffs) or customs in order to increase the price of imports and so protect domestic production

trailhead attraction A primary attraction to which visitors flock, from which they can be redirected to secondary attractions benefitting the wider district, city or region

trans-boundary pollution Pollution which crosses national boundaries

transhumance Animals are tended on upland pastures during the summer, while fodder crops are grown in the valleys; the animals are then moved down for winter stall feeding

transnational corporations (companies) (TNCs) Corporations or companies that operate in at least two countries

trickle-down effects The diffusion of the benefits of urbanisation, such as economic growth and prosperity, to poorer districts and people

undernourishment Measure of hunger referring to not having enough food to develop or function normally

underpopulation A population that is too small to fully utilise its resources; or a situation in which the resources could support a larger population without any reduction in living standards

United Nations Environment Programme (UNEP) The leading global environmental authority and advocate. It arguably sets the global environmental agenda, and seeks to promote the coherent implementation of the environmental dimension of sustainable development (within the United Nations system).

urban resurgence Population movement from rural back to urban areas, such as university students and upwardly mobile young people, reviving inner city and CBD areas

urban sprawl Unplanned growth of urban areas into the surrounding rural or rural-urban fringe areas

vectors Living organisms that can transmit infectious diseases between humans, or from animals to humans; they cause more than 1 million deaths annually

vertical integration An industry where one company either owns or controls multiple stages in the production and distribution chain

virtual water The volume of freshwater used to produce a product, measured at the place where the product was actually made

waste stream The complete flow of waste from its source through to recovery, recycling or disposal

water catchment An area of land through which water from any form of precipitation drains into a body of water (including groundwater supplies)

water scarcity Severe water stress; it is largely accepted that this occurs when annual water supplies fall below 1000 m^3 per person

water security The ability of a country to protect access to safe water resources for all the population

water stress When the demand for water exceeds available water resources, or when poor quality restricts its use

world cities Interconnected global economic centres that have significant influence on the world economy

World Trade Organisation (WTO) A global organisation that deals with the rules of trade between nations

zone in transition An area immediately surrounding the city centre of mixed land use – including older industries, terraced housing and areas of improvement and redevelopment